Henry Nicholas Ridley

The Natural History of the Island of Fernando de Noronha

Henry Nicholas Ridley

The Natural History of the Island of Fernando de Noronha

ISBN/EAN: 9783337329600

Printed in Europe, USA, Canada, Australia, Japan

Cover: Foto ©berggeist007 / pixelio.de

More available books at **www.hansebooks.com**

THE

NATURAL HISTORY

OF THE ISLAND OF

FERNANDO DE NORONHA

BASED ON THE COLLECTIONS MADE BY

THE BRITISH MUSEUM EXPEDITION

in 1887,

FROM

THE JOURNAL

OF

THE LINNEAN SOCIETY,

1890.

[*Extracted from the* LINNEAN SOCIETY'S JOURNAL—BOTANY, vol. xxvii.]

NOTES ON THE BOTANY OF FERNANDO NORONHA.

By H. N. RIDLEY, M.A., F.L.S.

[Read 7th June, 1888.]

(PLATES I.-IV.)

INTRODUCTION.

ON July 9th, 1887, the writer, with Mr. G. A. Ramage, of Edinburgh, started for Brazil to thoroughly explore the island of Fernando Noronha, lying in long. 32° 25′ 30″ W. and lat. 3° 50′ 10″ S., at a distance of 194 miles N.E. from Cape San Roque, coast of Brazil. On arriving at Pernambuco we were joined by the Rev. T. S. Lea, who came as a volunteer at his own expense. The cost of the expedition was defrayed by the Royal Society. After some delay at Pernambuco we embarked in the 'Nasmyth' steamship, trading to Liverpool, which was permitted to land us at the island, as the regular steamer trading between Pernambuco and Fernando Noronha was delayed for a long time just as she was due to start. We arrived at our destination on August 14th, and remained there till September 24th, when we returned by the little Brazilian steamer to the mainland. During our stay we were very hospitably entertained by the Capitao J. A. F. de Mendonça, the Governor of the island, who provided us with board, lodging, and horses, and a very intelligent and useful negro convict as a guide. We occupied ourselves in exploring and collecting plants, animals, and rock-specimens in all parts of the main islands, and visited also most of the other islets which were accessible; but owing to the absence of boats, which are not permitted on the island, we were unable to obtain much by dredging. The coral-reefs, however, at low tide afforded an abundant harvest of marine animals and plants. When we arrived at the islands the rains had just ceased, and the herbaceous plants were in flower; as we left the dry season was commencing, and the herbs were withering and the trees and shrubs were beginning to flower. Several of these latter, indeed, only commenced to flower just a day or two before we left; so that we were only able to procure a few flowers, and in two instances only flowers of one sex, and the fruits also were unprocurable. This latter defect, however, was in some measure made up for by the kindness of the Director, who after our return sent a box of carefully labelled fruits and seeds of some of the rarer plants. The nesting-season of the birds had just begun, and we were able to procure nests of two of the endemic land-birds. The sea-birds apparently nested somewhat earlier, as we found young birds almost ready to fly of several species. The insects were tolerably plentiful; but we were rather late for Lepidoptera, and the Coleoptera also seemed not to be at their best.

During our visit we had only one or two wet days, and usually the sky was bright and clear, with a strong breeze from the south-east. During December, January, and February we were informed that the island becomes very dry. Most of the streams and puddles are dried up and water is scarce. All the herbaceous plants of the central districts wither, and are set fire to, so as to clear the ground.

GENERAL ACCOUNT OF THE GROUP.

The whole group of islands forms a chain about eight miles in length, and probably at no very distant date were all connected. Indeed there is very good evidence to prove that the whole was at one time of much larger extent (see p. 17, note, and the Geological Report, pp. 86–94).

The largest island is the main one, about five miles in length and nearly two miles across in one spot, viz. near Tangle Rock, but otherwise very much narrower. The next in size is Rat

Island (Ilha dos Ratos), about a mile long, the most easterly of the group. Next to it, and apparently comparatively recently separated from it, is Ilha do Meio, or Booby Island. Then follows Sella Giucta, called in the Admiralty chart St. Michael's Mount*, a large phonolite peak rising straight from the sea; and between that and the main island is a low, flat, coral-reef island like that of Ilha do Meio, called Ilha Raza, or Egg Island, a little north of which are a few rocks forming a connexion with San José, or Platform Island, on which are the remains of a fort. On the south of the main island lie, at intervals, several rocky islets almost entirely barren of vegetation; and on the north side are two very similar conical basalt rocks known as Dois Irmaos.

Rat Island is a basaltic island of some size, the eastern end of which terminates in lofty crags, the haunt of numerous sea-fowl. The cliffs are lower on the north and south-eastern sides, and the ground slopes away to the west, where the basalt is overlaid by a considerable deposit of coral-reef, which, again, is covered with a layer of guano. At one point on the south-east corner the waves have eroded the reef so as to form a blowhole through which the spray rushes with great violence, so that the fountain can be seen at a distance of five or six miles. Round this blowhole was found *Sesuvium distylum*, n. sp., forming bright yellowish-green patches. The guano-ground was covered with a thick growth of *Ipomœa Batatas*, *I. pentaphylla*, *Phyllanthus*, *Momordica Charantia*, *Phaseolus lunatus*, *Ricinus communis*. Further inland the chief vegetation consisted of *Scoparia dulcis*, *Cyperus ligularis*, *C. brunneus*, *Æschynomene hispida*; while on the cliffs were the usual cliff-flora of this district— *Canavalia obtusifolia*, *Philoxerus vermicularis*, *Cereus insularis*, &c.

The only plants found here, and nowhere else in the island, were *Scoparia purpurea*, n. sp., *Sesuvium distylum*, *Cenchrus viridis*.

Owing probably to want of shelter from the winds there are no trees on the island, the Fig, *Ficus Noronhæ*, being reduced to a large shrub. There were a considerable number of weeds introduced by man, due partly to the settlement of Capt. Roma

* Most of the names on the Admiralty chart given by the French and English expeditions sent to the island are utterly unknown to the inhabitants of the island, whose nomenclature I have preferred in this report.

in the island for the purpose of working the guano. But the island has been inhabited to a certain extent at some time, as it is said that refractory convicts were formerly turned loose on the island by way of punishment.

There is a good permanent spring of water on the north side of the isle. The fauna includes, besides sea-birds, a Dove, *Zenaida Noronhæ*; but the Tyrant and *Vireo* found on Fernando Noronha are here absent. A Lizard and an *Amphisbæna* are both very abundant. Insects were very plentiful, though only a few species were taken ; and one of the endemic mollusca was plentiful on the west coast.

The next island is Ilha do Meio, and, like the adjacent portion of Rat Island, it consists of a thick deposit of coral-reef overlying the basalt. As mentioned above, it is evidently only a detached portion of Rat Island, and probably but recently separated. The surface is very flat, so that no bushes or trees can grow upon it, excepting that upon the cliff-faces Cacti, *Oxalis Noronhæ*, and other plants exist, being protected from the wind.

The reef is worn into holes and caves, and the low cliffs shelter many nesting sea-birds, whence its name of Booby Island.

Sella Giueta, or St. Michael's Mount, is a peak of phonolite rising almost vertically from the sea, and, owing to the violence of the surf beating on it, is very difficult of access. It was visited by Professor Moseley during the 'Challenger' expedition, and is covered, where the surface of the rock permits it, with an abundant native flora, including *Sapium, Capparis Cynophallophora, Cereus insularis, Oxalis Noronhæ, Dactylæna micrantha*. As it has never been inhabited, all the weeds of cultivation except *Amaranthus* are absent ; but the Lizard is abundant and large, and tamer here than on the other islands. A few insects occurred, and the Dove was plentiful, but neither the *Vireo* nor the Tyrant. Several of the sea-birds nest here, including the Tropic-bird (*Phaethon æthereus*) and the Frigate-bird (*Fregata aquila*).

From the fauna and flora of this spot it appears that the island was stocked while still connected with the other islands.

The islands lying between the Sella Giueta and the main isle call for little remark. The biggest is Ilha Raza, or Egg Island. It much resembles Ilha do Meio in form and structure, and the flora is similar. San José, or Platform Island, is connected with the main island by a ridge of basalt-rock only exposed at very low tides, as indeed is the preceding island. An old ruined fort surmounts Platform Isle, in the ruins of which we found numerous plants of *Solanum paniculatum*, and *Ipomœa Tuba*, while other species were common. A weak hairy form of *Eleusine œgyptiaca* occurred, and a single specimen of a new species of *Pupa* was found beneath a stone. The island is remarkable from the fact that the coral-reef here is much higher than that of the adjacent islands, 95 feet above sea-level. On the summit of the island were some large blocks of sandstone, apparently formed of blown sand containing shells.

The main island is long and narrow in outline, about five miles in length, and nearly two in breadth at Tobacco Point, where it is broadest. The centre of the island forms an undulating plateau about 200 feet above sea-level, sloping upwards at the western end, and terminating in a long, narrow, inaccessible promontory known as Cape Placellière. The cliffs are high and often perpendicular, sometimes descending into the sea, but often with sandy bays at the foot. At the eastern end of the island the ground slopes away to sea-level, and here are extensive sand-hills covered with *Ipomœa Pes-capræ, Pavonia cancellata,* and *Sida altheæfolia.* The soil of the central district is a fertile red clay, formed by disintegration of the basalt which forms the bulk of the island. This portion is mostly under cultivation, and the flora consists for the greater part of introduced weeds, but here and there are a few endemic plants. The hills are cultivated also almost to the summits here; but upon the East Hills, the Peak, and Tangle Rock were obtained a number of native species, growing mingled with weeds of cultivation. Two species of plants, viz. *Combretum* sp. and *Aspilia Ramagii,* were only found on the East Hills; while the Peak and Tangle Rock, both phonolite rocks, produced several endemic species. The western end of the island is covered with dense forest, but large trees are now not common, owing to the demand for firewood and to the strict orders for the destruction of all large trees to prevent the convicts making rafts of them on which to escape. This portion of the island is termed the Sapate; the chief trees and shrubs there are *Sapium sceleratum, Bignonia* sp., *Schmidelia insulana, Jacquinia armillaris, Oxalis Noronhæ, Bumelia fragrans,* n. sp., *Anacardium occidentale, Palicourea* sp., *Pisonia Darwini, Spondias purpurea, Jatropha Pohliana, Capparis Cynophallophora, C. frondosa,* and *Croton* sp. Except along the paths in the wood there is very little undergrowth in the Sapate ; but a new species of *Oxalis* was obtained at one spot. At the entrance to this district, along the path used by the woodcutters, a number of weeds occurred, not common elsewhere, which were no doubt introduced by the wood-cutters themselves, such as *Plumbago scandens* and one or two large patches of *Panicum numidianum*. The latter is the culti-vated fodder-grass of Brazil, and from its position here it appeared to have been brought in the form of hay for the horses used in carrying the wood from the forests.

There is a large pool of water of considerable depth on the south side of the Sapate ; it is surrounded at some distance by a semicircle of high cliffs, between which and the lake is a dense thicket of shrubs, which come down to the very edge. The lake is fringed with *Panicum brizoides* and almost filled to the brim with *Nitella cernua* and an Alga, among which we found many specimens of a new species of *Planorbis* and several aquatic insects not met with elsewhere. On Morro branco, a hill com-posed of phonolite altered by contact with basalt, a few local plants were found, and a peculiar *Paspalum* with stiff erect leaves (*P. phonoliticum*). Wherever the cliffs were broken up they were covered with a vegetation of maritime plants, such as *Canavalia obtusifolia, Philoxerus vermicularis, Ipomœa Tuba, I. Pes-capræ, Cenchrus echinatus*, &c.

HISTORY.

The island was first discovered by Amerigo Vespucci in 1503, in his fourth voyage. A fleet of vessels having been despatched from Spain under Coelho, sailed first to the Canary Islands, then to Sierra Leone, and thence attempted to reach Bahia, which had been discovered during a previous voyage in 1501. The discoverer published his account of the finding of the island in the ' Lettera di Amerigo Vespucci delle Isole nuovamente trovate in quattro suoi viaggi,' from a translation of which (Quaritch, 1885, p. 43) I take the following account :—" And when we had sailed full 300 leagues through the immensity of the sea, being then quite 3 degrees south of the equinoctial line, we became aware of a land from which we were probably 22 leagues distant: whereat we marvelled : and we found that it was an island in the middle of the sea and was very lofty, a very marvellous work of nature : since it was no more than two leagues in length and one in breadth : in which island never had there been in habitation by any people, and it was Bad Island for all the fleet, for your Magnificence must know that by the ill-counsel and steering of our Chief Captain he lost his ship here: since he struck with it upon a rock, and it split open on St. Laurence's night, which was on the 10th day of August, and went to the bottom: and there was nothing saved thereof except the crew. It was a ship of 300 tons; in which went all the importance of the fleet; and when all the fleet had laboured to save it, the Captain commanded me to make with my ship for the said island to seek a good anchorage where all the ships might anchor: and as my boat manned with 9 of my sailors was in service and aiding to belay the ships, he willed that I should not take it and that I should proceed without it: telling me that they should take it to me to the island. I quitted the fleet for the island as he ordered me, without a boat, and with the deficiency of half my crew, and I went to the said island, which was about 4 leagues distant: in which I found an excellent harbour, where all the ships could anchor very safely: where I awaited my Captain and the fleet fully 8 days, and they never came: so that we were very discontented, and the men that had remained with me in the ship were in such dread, that I was unable to console them; and being thus, the eighth day we beheld a ship coming upon the sea, and from fear that it might not see us, we weighed with our ships and made for it, thinking that it brought me my boat and crew." However, the rest of the fleet with the boat had gone further south (p. 44); so "We returned to the island and provided ourselves with water and timber by means of my companion's boat, which island we found uninhabited, and it contained many fresh and sweet waters, innumerable trees full of so many sea- and land-birds that they were beyond count, and they were so tame that they allowed themselves to be taken with the hand, and so many of them did we take that we loaded a boat with those animals. We saw none (other) except very large rats and lizards with two tails and some snakes." ["Infinitissimi arbori plena di tanti uccelli marini e terrestri che eron senza numero, . . . et tanti ne pigliamo che carichamo un battello di epsi animali; nessuno non vedemo; salvo Topi molto grandi e Ramarri con due code et alchuna serpe."—*Lettera*, Fiorenza, 1505; Quaritch's Reprint, London, 1885 (unpaged).

They then made provision, and departed by the wind between S. and S.W. for Bahia, which they reached in seventeen days, and it was 200 leagues from the island.

Some geographers seem to have been doubtful as to what this island described by Vespucci was, and several other islands were suggested, including the mythical St. Matthew's Isle and St. Paul's Rocks; but the position assigned by Vespucci, and the presence of abundant fresh water and trees, negative this suggestion. Humboldt and most other geographers, however, seem to agree that this island was certainly Fernando Noronha.

There are several very interesting points in the account quoted above. First, the author merely mentions one island. Now, without doubt the whole chain was connected at one time, but whether or not it was so when Vespucci discovered it must remain doubtful. The wrecked ship was lost probably off Rat Island, the first point that they would come to; and if Vespucci anchored in San Antonio Bay, on the north side, which is the nearest good anchorage, he would, as he says he was, be unable to see the rest of the fleet, owing to the high ground of Rat Island between him and it. His description of the trees and innumerable birds is evidently correct, though most of the trees are destroyed, and the birds far less abundant than they were then. The lizards with two tails may have been a confusion of the very abundant and conspicuous Gecko with the *Amphisbæna*, which is often called the snake with two heads, or may have been suggested by finding an accidentally fork-tailed lizard, an example of which monstrosity was obtained by our expedition. The serpents were doubtless the *Amphisbæna*. But the large rats are much less easy to explain; at present the only rats occurring on the island are *Mus rattus*, the common introduced black rat. It is impossible that the animals seen by Vespucci could have been this species, which could not at that time have been introduced. Is it not probable that there was formerly an indigenous rat-like mammal, exterminated by the introduction of the black rat? We could find no tradition even of this big rat, and I fear it is quite extinct. The only hope of recovering its remains lies in the guano deposits of Rat Island, where its bones might be preserved.

ORIGIN OF THE FLORA.

Before suggesting an origin or origins of the present Flora of the group, it must first be pointed out that there is no evidence whatever to show a former connection with the mainland of Brazil at any time, in spite of what has been asserted by Dr. Rattray

(Quart. Journ. Geol. Soc. xxviii. p. 33) to the contrary. There are no sedimentary rocks on the island, and granite, the prevailing rock in the neighbourhood of Pernambuco, is entirely absent, while at the same time no basaltic or phonolitic rocks are known from the adjacent mainland*. The " Tertiary conglomerates " of Rattray, showing a former connection with the Tertiary rocks of the adjacent mainland, are quite mythical. In spite of all attempts, it seems quite impossible to fix certainly the period at which the island rose from the bed of the ocean. That it is of considerable antiquity there is little doubt, however. From the petrological structure of the island it seems certain that it rose from the bottom of the ocean at some remote period, and of course when this happened there was no vegetation upon it. How is the present Flora to be accounted for ? Most of the plants may be relegated to one of three classes :—Weeds, or plants introduced intentionally or accidentally by man ; plants of which the seeds or fruits are known to be carried about the ocean by currents ; and plants with eatable fruit which is sought by birds.

Weeds.

To this class belong many of the species. They include all the Malvaceæ and nearly all the Leguminosæ, the remainder being scattered over other orders. Most of them are plants of world-wide distribution and very common on the adjacent mainland. Few or none occur on the smaller islets, such as Sella Giueta ; but where there have been settlements these plants seem to spring up at once. By far the greater number were to be found on the main island in the open central district and in the village. Only one occurred on Sella Giueta, and that was *Euxolus viridis*. To this section belong all or nearly all the plants with adhesive fruits or seeds, viz.:—*Desmodium* (4 species), *Æschynomene, Zornia, Plumbago, Boerhaavia* (2 species), *Chloris* (2 species), *Eragrostis ciliaris, Anthephora, Cenchrus* (2 species), *Setaria scandens.*

The absence of these from the smaller islets seems to show that the bird-fauna is not responsible for their presence here ;

---

* When comparing Fernando Noronha with the adjacent continent, it is but just to point out that the nearest point, Cape San Roque, and indeed the whole of the Province of Rio Grande del Norte, is almost entirely unknown as regards its geology and natural history, and that even in the neighbourhood of Pernambuco much remains to be done.

for, as we shall see, the plants whose seeds are eaten by the birds,
and the *Gonolobus*, whose downy seeds line the nest of the Tyrant,
are carried about everywhere. But, on the other hand, where the
convicts had made paths through the woods, and especially where
they were able to take the horses for the purpose of fetching firewood, these weeds had followed and established themselves.

Plants introduced by Sea-currents.

Mr. W. B. Hemsley, in Bot. Voy. Chall. pt. iii. App. pp. 277–313,
has given a good account of all that is at present known about
these plants, but much yet remains to be ascertained. I
fear I can add but little, for though both at Pernambuco and
on Fernando Noronha we carefully sought for drift-fruits and
seeds, we were only rewarded by finding two seeds of *Mucuna
ureus* in Sueste Bay, a plant not yet established there. The
current which strikes the island of Fernando Noronha is one
which passes up along the east coast of Brazil. This current
would naturally strike the island on its south side, and would
bring with it seeds from the southern regions of Brazil. It
would be aided also at the time of year at which we were
upon the island by the trade-winds, which blow from the south-
east, and, indeed, we found upon the sands of the bays on
that side numerous marine plants and oceanic animals, such as
Velellas, *Physalias*, and *Ianthina*, which we did not see at all
on the northern side, besides the above-mentioned seeds of
Mucuna. But there are a number of plants upon the island of
which the seeds are known to be carried about the sea in this
way, having been met with in sea-drift. These are *Canavalia
obtusifolia, Rhynchosia minima, Abrus precatorius, Acacia Farnesiana, Ipomœa Tuba, I. Pes-capræ, I. pentaphylla?, Philoxerus
vermicularis, Talinum patens, Portulaca oleracea, Ricinus communis, Laguncularia racemosa.* Besides which, species of the
genera *Sesuvium*, *Erythrina*, and *Pisonia*, each of which here
supplies an endemic species, have been met with as drift-seeds.

Jatropha Pohliana and *J. urens*, both common on all parts
of the main island, and also on Rat Island, are probably also
drift-seeded plants. The bark and wood of the former was very
common on the shores at Pernambuco.

Canavalia and *Philoxerus* occurred all along the coasts of Rat
Island and the main island, and very rarely went very far from
the beach. *Acacia Farnesiana* may have been introduced by

man, as it did not grow away from cultivation; and I suspect also that *Portulaca* is here at least a weed of cultivation rather than a sea-drifted plant, with which Mr. Hemsley classes it. *Ricinus* is very widely spread over the islands, and is most probably introduced by sea-currents. The convicts used to say that wherever the ground was dug *Ricinus* used to come up. Very possibly some of the other hard-seeded Euphorbiaceæ, *Euphorbia comosa*, *E. hypericifolia*, which always grows on the shore, and *Croton odoratus*, were originally introduced by ocean-currents.

Ipomœa Tuba is interesting, as it is not known from Brazil south of Fernando Noronha, and is a native chiefly of the West Indies; and *Cyperus brunneus*, Sw., which is common here, has not been obtained anywhere out of the West Indies except Florida and Mexico, and is one of the few plants known from S. Trinidad, where it was obtained by Sir Joseph Hooker, and published under the name of *C. atlantica*, Hemsl.

Several species of plants which might be reasonably expected to have been drifted across do not occur, notably *Remirea maritima*, *Fimbristylis glomerata*, *Avicennia*, and *Conocarpus*, all common in the sands of the neighbouring mainland; and the Cocoa-nut appears also to have been a recent introduction by man, although the shores below Pernambuco are lined with groves of them.

Plants with Berries and Eatable Seeds.

To this group belong a large number of plants, including several endemic species. Two species of *Capparis*, several species of Cucurbitaceæ, including three species of *Ceratosanthes*, all endemic; two *Cayaponias*, and a *Momordica*, *Cereus*, *Palicourea*, *Guettarda*, *Bumelia*, *Physalis*, and *Ficus*, all endemic species; *Jacquinia*, *Vitis*, *Rauwolfia*, *Cordia*, *Rivina*, four *Solanums*. Besides these are several plants originally introduced by man intentionally, which are now scattered all over the main island by the birds: such are *Solanum oleraceum*, *Capsicum frutescens*, *Basella alba*, *Spondias purpurea*, *Anacardium occidentale*, *Carica Papaya*, and *Lycopersicon esculentum*. Now a considerable number of these are to be found on the smaller islands as well as in the most inaccessible spots of the main island. One *Capparis*, several Cucurbitaceæ, the *Cereus*, *Rivina*, and *Ficus* occur on Sella Giueta. The fig, indeed, grows in almost every spot at all suitable for it, even on the highest parts of the inaccessible portion of the Peak, on the isolated rocks called Dois Ismaos, and in many high inaccessible crags.

There is only one fruit-eating bird upon the island, and that is the endemic dove, *Zenaida Noronhæ*, which is exceedingly abundant, and flies from island to island. The crops of the specimens shot we frequently found full of the *Cayaponia* fruits.

When one sees the number of endemic species with edible fruit, one is tempted to wonder if it were possible that they were all introduced by this single species of Dove, or whether other frugivorous birds may not at times have wandered to the shores.

One is too apt to imagine that only gaily coloured berries are attractive to birds, and we were thus puzzled to account for the *Sapium* occurring so widely over all the isles and in very inaccessible spots high upon the rocks ; but we were informed by our guide that the small birds eat the seeds greedily and pass them uninjured, thus scattering them about the island. As the seeds are so poisonous that they are said to blister the skin of any horse or cow on which they fall, it is surprising to hear that the birds are fond of them.

The Relations of the Flora to the Insect Fauna.

It will be noticed that there are in the Flora a considerable number of plants which require the aid of insects in fertilization. The Cucurbitaceæ, the Papaw, *Schmidelia*, *Combretum*, *Terminaliopsis*, are all diœcious, and not being anemophilous, must be fertilized by insects. *Oxalis Noronhæ* also has dimorphic flowers, implying the necessity of insect-fertilization. Many plants have showy coloured flowers, the commonest colour being yellow. *Datura Stramonium*, *Cereus insularis*, and *Ipomœa Tuba* are nocturnal plants with white flowers, very sweetly scented in the first two cases at least. Several species, such as *Urena lobata*, *Oxalis Noronhæ*, and the cultivated Cucurbitaceæ, open their flowers in the early morning, closing them when the sun gets hot, about 10 o'clock ; *Palicourea* and *Bumelia* seem to be really diurnal, and are strongly scented during the daytime. All the plants above mentioned fruited very extensively on the islands, and the Leguminosæ and Cucurbitaceæ were especially productive.

The number of insects belonging to the orders which are well known as plant-fertilizers is surprisingly limited. A few small species of moths haunted at night the bushes of *Scoparia dulcis, Cassias,* &c. on the open spaces. A single species of butterfly was very abundant on Rat and the main island, but we never saw it visiting flowers.

The most important fertilizer was a small endemic hornet belonging to the genus *Polistes,* which gathered honey from the Leguminosæ and Cucurbitaceæ; and three small black species of *Halictus* were caught in the flowers of the melons, *Momordica Charantia, Oxalis Noronhæ,* and the mustard. The latter plant was also haunted by *Temnoceras vesiculosus,* a pollen-eating Syrphid. The only other insects which could also be considered as possible fertilizers were *Tachytes inconspicuus,* n. sp., and *Monedula signata,* two sand-wasps, *Pompilus nesophila,* n. sp. (Hymenoptera), and *Psilopus metallifer* (a Dipteron), but none of these were seen at or near flowers. A small black beetle also was found in the flowers of an *Acacia* in the Governor's garden.

Though the number of species of insects was not large, yet the individuals, especially of the *Polistes* and *Halicti,* were very numerous, but at the same time they seemed out of all proportion to the immense number of flowers to be fertilized. It is very probable, however, that the majority of the Leguminosæ and some of the other plants were self-fertilized.

Groups of Plants rare or unrepresented.

The absence of plants or groups of plants from a given locality often throws as much light on the origin of the flora as the presence of others does. In the present case the absence of marsh-plants is among the most striking; for in the first place they are exceedingly abundant on the adjacent mainland, and, again, there is every reason for their being introduced by the wading-birds which fly across from Brazil. The genera *Eleocharis, Utricularia, Pœpalanthus, Saleria,* and many others remarkably abundant on the adjacent mainland, are here quite absent. Indeed, the only really marsh-loving plants met with were *Jussieua linifolia, Ammania latifolia,* and *Panicum brizoides.* The dryness of the island during the dry season accounts for this in the main; but there are spots which are permanently damp, and here one might reasonably expect to find marsh-plants. The chief plant which grows along the streams and on these damp spots is, however, *Philoxerus vermicularis*; and as the water has a brackish taste, it is probable that the salt or other mineral matters in the soil prevent the growth of purely marsh-plants. The dryness of the climate is also no doubt the reason for the absence of sylvestral plants, the Sapato woods, where such plants would naturally occur, being dry and rocky.

The absence of petaloid Monocotyledons from oceanic islands has been commented on by Hemsley, Chall. Exp. Report, Bot. vol. i., and Fernando Noronha is no exception to the rule. Plants with winged or feathery seeds are supposed to possess great faci-lities for being widely disseminated. That this is really the case may be doubted. The only species with winged or plumed seeds here are *Gonolobus micranthus, Jussieua linifolia,* and *Ageratum conyzoides.* The first of these is endemic; its plumed seeds are used by the endemic Tyrant, *Elainea Ridleyana,* to line its nest with. Is it conceivable that the seeds of the ancestor of this plant were accidentally brought over attached to the feathers of some bird which had, in like manner, used them for its nest ? *Jussieua* is a marsh-plant, with small seeds plumed like those of an *Epi-lobium*; it and the *Ageratum* only occurred in the cultivated ground in the centre of the island. The latter certainly, and possibly the former, were introduced as weeds by man.

Bromeliaceæ, though commonly provided with plumed seeds and abundant in Brazil, are quite absent here. In reality it would require a strong and very long-continued wind to carry plumed seeds to such a distance from the land, and even if such should be the case, the chances of seeds dropping upon a small island like this would be exceedingly remote.

Recent Alterations in the Flora.

When the island was first discovered in 1503, Vespucci found there infinite numbers of trees, most of which have now dis-appeared. Of *Erythrina exaltata,* mentioned by Webster as the largest tree in the island, only one full-sized tree now remains ; and as it seems that the young trees will not flower, the species appears to be threatened with speedy extinction. Of the fig-tree, again, but few large ones remain, the finest being in the Governor's garden. This is due to an order of the govern-ment, which provides that all trees of a sufficient size to be made into rafts shall be destroyed, for fear that the convicts might escape on them ; and besides this the constant demand for fire-wood causes great destruction amongst the smaller shrubs. I could not find, however, from the inhabitants, that any perceptible diminution in the number of trees had occurred of late years, neither do the more recent accounts of Webster, Darwin, or Mosely lead one to suppose that the island was very much richer in trees than it is now.

Another cause for the change of the flora is to be seen in the more recently introduced plants, and the creeping and climbing plants seem to be rapidly destroying the older vegetation. The number of climbing plants on the island is very large, belonging chiefly to the orders Cucurbitaceæ and Leguminosæ. The former, especially *Momordica Charantia* and *Cayaponia Tajuga*, cover the trees and bushes on the edges of the forest with a dense mat of stems, so that they are soon suffocated and destroyed, and when they have fallen to the ground they are soon covered with a carpet of thickly woven stems of Cucurbitaceæ and Leguminosæ, of which *Phaseolus peduncularis* appears to be the most destructive, and on the ground thus once covered the shrubs can no longer reassert themselves. Furthermore during the dry season, December to February, these climbing-plants wither and dry up and are set on fire, and any seedlings of the shrubs which may have escaped strangulation by the climbers are destroyed. The conflict between the climbing-plants and the shrubs was very well seen all along the eastern edge of the Sapate. In the woods themselves these plants were entirely absent, since they were unable to grow among the dense shrubs, from want of light and air.

The Freshwater Fauna and Flora.

The number of permanent streams and pools in the whole Archipelago is very small, as in the dry season almost all dry up. There is a spring on Rat Island said to be never dry, and there are also one or two on the main island, where besides there is the largest stream at Sueste, and also the lake in the south-west corner, in both of which aquatic plants and animals might occur; none of these, however, are rich in fauna or flora. The lake contained a species of *Nitella* and an alga, an aquatic beetle and Hemipteron, a new species of *Planorbis*, and an Ostracod, the latter also occurring in all the streams of any size. The remaining streams and puddles produced dragonflies, a species of *Gammarus*, and a few algæ. One may compare this with the freshwater fauna and flora of the other Atlantic islands. The absence of freshwater fish and amphibians is common to most small islands.

Freshwater mollusks occur in several islands, including Madeira, where are a species of *Ancylus* and *Lymnæa*. The Azores possess no freshwater mollusk according to Godman, who attributes this to the paucity of waders and ducks inhabiting these islands, although he gives a very considerable list of swamp-loving birds as occasional migrants. In the Galapagos Islands a *Paludina* occurs.

Water-beetles are apparently always very rare in oceanic islands. Wollaston shows that the *Hydradephaga* are the most poorly represented group in the Atlantic islands visited by him (Coleopt. Atlantidum, Introd. p. xv). In St. Helena, too, there are none; and in Fernando Noronha they seem scanty.

DISTRIBUTION OF THE FAUNA COMPARED WITH THAT OF THE FLORA.

It is unfortunate that the naturalists who have visited oceanic or distant islands have usually examined into the distribution of one group, either plants or animals, or often but one order of the latter, so that it is very difficult to obtain any clear ideas as to the relations of the two groups. I have had in presenting these reports the assistance of my colleagues in the British Museum, and other English naturalists of the highest standing, and am therefore able to make a few observations on the distribution of both plants and animals as compared together.

Just as in plants, we have a considerable number of animals introduced by man into the islands intentionally and by accident: such, for instance, are the Gecko (*Hemidactylus mabouia*), the American Cockroach (*Blatta americana*), and its curious parasite *Evania*, a spider, centipede scorpion, rats and mice, *Sitoplidus oryzæ*. These, though usually plentiful on the main island around the houses, are markedly wanting from the smaller islets.

There is also a large group which has arrived here by the aid of their wings, probably assisted by a suitable wind. This includes a number of the peculiar terrestrial fauna, the landbirds and the insects. In looking over the lists of species taken here, we may note that the smaller birds are endemic, and a large proportion of the smaller insects. The small butterfly and almost all the moths are known from the mainland of South America, and the dragonflies are also widely distributed forms. All the winged fauna have a South-American facies, whether they are endemic or of wider distribution.

There is another group which is unprovided with means of traversing the ocean, and not carried about by man. This includes the *Amphisbæna*, Skink, the freshwater and terrestrial Mollusca, and perhaps some of the feebler-winged and apterous insects, the endemic ostracod, &c.

The *Planorbis, Gammarus*, and Ostracod, all (?) endemic species, it is quite conceivable may have been brought over on the feet of Waders, which seem to migrate here.

The remainder are more difficult to account for. The Mollusca are almost all peculiar, and the two that are not so are West-Indian. The *Amphisbæna* and Skink are endemic, and allied not to Brazilian but to West-Indian forms.

It is commonly said that reptiles and terrestrial mollusks find their way across the ocean by secreting themselves or their eggs in floating trees, which are drifted to islands; and though for several reasons this does not seem a satisfactory explanation of their distribution, yet the distribution of these animals here points to this as the means by which they have arrived. As I have said, they are West-Indian in facies, and correlated with this is the striking fact that the marine fauna and flora, and at least one of the plants whose seeds are known, supposed to be constantly drifted about the sea, and to be thus carried from place to place, is only known also from the West Indies (*Ipomœa Tuba*). Another fact of interest in connection with these sea-travelling fauna, if I may use the expression, is the fact that almost all occur on all the islands suited for their existence. Thus, on Rat Island the *Bulimus Ridleyi*, the *Amphisbæna*, and Skink are common on St. Michael's Mount; the Skink is a large species, but the island, being a mere rocky peak, is unsuited for the *Amphisbæna*.

On Platform Island the lizard and several terrestrial Mollusca were found, while at the same time almost all the animals of a more recent introduction were absent from these localities, just as is the case in the distribution of the plants. I believe, in fact, that this part of the fauna and flora was established on the island before it was broken up into the little archipelago of rocks and islets of which Fernando Noronha now consists *. Perhaps

---

* On reference to A. Vespucci's description of the place, it will be found that he speaks of it as one island, so the breaking-up into an archipelago may only have taken place within the last 400 years.

even this portion of the fauna and flora was introduced previously to the deposition of the basalt over the masses of phonolite, which form as it were the skeleton outline of the island.

I cannot find any recorded observations of the flow of a current in the sea now from the direction of the West Indies; on the contrary, the current marked on the maps, and which was certainly during our visit throwing up fleets of *Velellas*, *Physalias*, Algæ, and other marine drift, was flowing from the south. No *Physalias* or *Velellas*, nor anything of the kind was to be found on the north side of the group, although the former at least were very plentiful in the open sea to the north. It is true that we did find some pieces of rotten timber on the north side of the island at the foot of the cliffs of the Sapate, but they may have been, and I think were, portions of the mast of a ship. They were buried under débris from the cliff and quite decomposed. And, again, I should add that during our visit the wind blew from the south, while we were informed that at other seasons it blew strongly from the north.

Published Accounts of the Island.

The earliest account is that by Amerigo Vespucci, mentioned above. The next which I have been able to find embodying any notes on natural history is that published by Juan and Ulloa in their Voyage to South America. These travellers arrived there on their way north from Cape Horn on May 21, 1744. They describe the island as very barren, from want of rain, saying that previous to their visit there had been no rain for two years. However, it must be remembered that the time of their visit was not long after the end of the normal dry season. They mention abundance of fish, including lampreys and morenos (*Muræna*), and describe a fish called a cope.

In the account of the voyage of the 'Chanticleer,' under Capt. Henry Foster, Mr. Webster, in vol. ii. pp. 326–339 of his narrative of the Voyage, gives a very good account of the geology, and some remarks on the botany and zoology. Even at that time there were few large trees on the island, the commonest being the "Bara." He mentions the *Jatropha*, *Cassia occidentalis*, *C. falcata*, and several species of *Indigofera*. The largest trees on the island were the "*Erythrina exaltata*." "The Acacias are the graces of the woods, and cast a sweet perfume around." By these I conclude he alludes to *Acacia Farnesiana*, but it does not occur in the woods at present. He also talks of the "*Swartzea pennata*, Jajo," which was possibly *Swartzia pinnata*, Willd., sometimes cultivated in Brazil.

The animals collected by this expedition are in the British Museum. They include the skink, dove, tyrant, and also *Thysanodactylus lineatus*, a large lizard not known now to occur in the island; but as the expedition collected at other places, it is quite possible that this specimen may have been collected elsewhere and mislabelled. Mr. Webster remained on the island a month.

The 'Beagle' landed its crew on the island on Feb. 20, 1832, and Mr. Darwin visited the Peak and made geological notes upon it, and also collected a number of plants, now in the Cambridge Museum. He says "the whole island is covered with wood, but from the dryness of the climate there is no appearance of luxuriance."

In 1871, H.M.S 'Bristol' visited the island to take some altitudes, and Dr. Rattray published an account of the geology in the Geological Society's Quarterly Journal, vol. xxviii. pp. 31–34, and a popular account of the island in the Geographical Society's Journal. Of these it will be sufficient to say that the geological account and map are erroneous and misleading—the phonolite being spoken of throughout as granite, and the so-called granite being marked on spots where basalt only occurs, while "tertiary conglomerates" are recorded from various spots and correlated with those on the mainland of Brazil. What was intended by these "tertiary conglomerates" is not clear, but probably masses of basaltic beach-pebbles cemented together by gypsum, which occur in some at least of the spots where the tertiary conglomerates were found.

In Sept. 1873, H.M.S. 'Challenger' arrived at the island with intention of exploring it, but being unarmed with the requisite authority, were refused permission. Prof. Moseley, however, succeeded in obtaining a few plants, both from the main island and from St. Michael's Mount, which were described and figured in the Voyage of the 'Challenger' Report, Botany, pt. ii., by Mr. Hemsley*. The officers of the 'Challenger' also took soundings at various distances from the ship, and dredged at some little distance from it. The animals obtained are in the British Museum; the plants were divided between that institu-

---

* For an account of this visit, see Journ. Linn. Soc. xiv. (1875), pp. 359–362.

tion and the Kew Herbarium. From the published descriptions there seems to have been little alteration since the time of Ulloa's visit in the appearance of the island.

Besides the specimens of plants and animals mentioned above as having been collected by Webster, Darwin, and Moseley, there is a small early collection of seven specimens of plants in the British Museum by an unknown collector, apparently a foreigner, and a single specimen of *Capparis Cynophallophora*, from Capt. Middleton, at Kew.

SUMMARY.

The whole group of islands possesses certain characteristics common to all truly oceanic islands, and some of those which are merely the relics of vanished continents. In the first place, there is the absence of indigenous mammals, and more noticeably of bats, of freshwater fish, and amphibians. Again, the number of indigenous species, both of plants and animals, is very small, while the number of individuals is very large. The insects are small and dull in colour, and but few of the plants have showy flowers, white and yellow being prevailing colours. A considerable proportion of the indigenous plants are shrubby or arboreous, as in many other oceanic islands; but arboreous or even shrubby Compositæ do not exist, indigenous species of the group being rare in the islands.

POLYPETALÆ.

CAPPARIDEÆ.

CAPPARIS CYNOPHALLOPHORA, *L. Sp. Pl.* p. 721; *Jacq. Am. Pl.* p. 158; *Eichl. in Mart. Fl. Bras.* xiii. 1. p. 282.

A common shrub on the main island, especially in the Sapate and near Tobacco Point. It is also occurs on Sella Giueta. It attains a height of about 12 feet; but in open spots is much smaller. During our visit it was hardly in flower, very many plants showing no signs even of buds. One fruiting specimen occurred; but from the number of seedlings in some spots, there can be little doubt it fruits extensively at some seasons. The flowers are white and fugacious. The fruit a soft pulpy red pod. The plant is much infested with galls. It is called " Feijao de lenha."

It was also obtained by Moseley and by Middleton.

Distribution. Florida, Panama, Yucatan, Mexico, most of the West-Indian islands, Guayaquil, Venezuela, and Brazil as far south as Rio de Janeiro.

CAPPARIS FRONDOSA, *Jacq. Am. Pl.* p. 162, t. 104; *H. B. K. Nov. Gen. et Sp.* v. p. 91; *Eichl. in Mart. Fl. Bras.* xiii. 1. p. 280.

A large leafy shrub, very abundant in the woods around the lake and in the open parts of the Sapate. It is used for making hoops for barrels, &c. It is about 12 feet in height, and would probably become a larger plant, but is much cut for firewood. The flowers are very fugacious; the petals dull purplish green, the stamens white; the fruit resembles that of the preceding species, but is larger, and at first green, then purple-rose, and finally black; it never appeared to show any signs of dehiscing into two valves, as is represented in *C. flexuosa*, Vell. Fl. Flum. v. t. 108; but the fruit as it ripened became soft and pulpy. Mice are very fond of the seeds.

Native name " Gito."

Distribution. West Indies and Brazil.

CLEOME SPINOSA, *L. Sp. Pl.* p. 939; *Ait. Hort. Kew.* ed. 2, iv. p. 131; *Eichl. in Mart. Fl. Bras.* xiii. 1. p. 253.—C. pungens, *Willd. Hort. Berol.* t. 18.

A form with thorns and white petals and pink stamens. Occurred in the garden of the Residency and in waste places in the village. As it is used in medicine, it was doubtless introduced to the island by man. It is found in many places in Brazil in a half-wild state; and we met with almost the same form in waste places in the town of Olinda, Pernambuco.

C. DIFFUSA, *DC. Prodr.* i. p. 241; *Eichl. in Mart. Fl. Bras.* xiii. 1. p. 258.

Very plentiful at one spot upon the upper part of the Peak, growing among Cucurbitaceæ. Flowers white.

Distribution. Brazil.

DACTYLÆNA MICRANTHA, *Schrad. Hort. Goett.*; *Schult. f. in Roem. et Schult. Syst.* vii. p. 9; *Eichl. in Mart. Fl. Bras.* xiii. 1. p. 243.—Cleome monandra, *DC. Pl. Rar. Hort. Gen.* p. 54, t. 15.

Abundant along pathways through the Sapate, and also on Sella Giueta. Flowers pink.

Distribution. North Brazil, Bahia, and Pernambuco.

CRUCIFERÆ.

The Cabbage, *Brassica oleracea*, L., and Mustard, *Brassica alba*, Boiss., are both very successfully cultivated here, but not in any quantity.

ANONACEÆ.

Anona squamosa, *L. Sp. Pl. ed. Willd.* ii. p. 1265. no. 3.

There are several trees of this species on the island, at Samba-quichaba and Sueste. They fruit well.

PORTULACACEÆ.

Portulaca oleracea, *L. Sp. Pl.* p. 638; *Haw. Misc.* p. 126; *Rohrb. in Mart. Fl. Bras.* xiv. 2. p. 229.

Very common among the stones in the village, and also on the sea-shore on the north side, both of the main island and Rat Island.

Distribution. All over the warmer parts of the world.

Talinum patens, *Willd. Sp. Pl.* ii. p. 863; *Rohrb. in Mart. Fl. Bras.* xiv. 2. p. 296.—Portulaca patens, *Jacq. Hort. Vindob.* ii. t. 151.

Common on the main island in thickets near the sea. A specimen was found in Portuguese Bay nearly 6 feet in height. It is also abundant on Rat Island and St. Michael's Mount.

The flowers are usually pink; but white-flowered plants occur also.

Distribution. All Central and South America.

MALVACEÆ.

Pavonia cancellata, *Cav. Diss.* iii. p. 135; *DC. Prodr.* i. p. 444.

Plentiful on the sand-hills at San Antonio Bay.

Distribution. South America from Surinam, and Caracas.

Hibiscus esculentus, *L. Sp. Pl.* p. 980; *DC. Prodr.* i. p. 450.

Extensively cultivated as a vegetable, as elsewhere, in Brazil.

Gossypium barbadense, *L.*

The cotton grown here is of very fine quality; but it is but little cultivated.

Urena lobata, *L. Sp. Pl.* p. 974; *DC. Prodr.* i. p. 441.

A patch of plants of this species occurred among *Cassia* and *Crotalaria* on the hill above the garden of the Residency. The showy pink flowers closed before 10 a.m.

Distribution. Throughout the tropics of both hemispheres.

Wissadula hirsuta, *Presl, Rel. Hænk.* p. 118; *Walp. Rep.* i. p. 328.

About a dozen plants grew among the bushes at the entrance to the Sapate woods. The flowers are yellow and rather showy. The plants were about 5 feet high.

Distribution. Brazil.

Malachra capitata, *L. Syst.* ed. xii. p. 518; *DC. Prodr.* i. p. 440; *Griseb. Fl. Brit. W. Ind.* p. 80.

This was plentiful in the central district, growing with the Cassias, &c., in the Horta da Florestas. It is a coarse half-shrubby plant with white flowers. This is the species mentioned under the name of *M. radiata* by Hemsley in the Botany of the ' Challenger' Voyage, Atlantic Islands, p. 15.

Distribution. Most tropical countries.

Sida altheæfolia, *Sw. Prodr.* p. 101; *Sw. Fl. Ind. Occ.* ii. p. 1207; *DC. Prodr.* i. p. 464.

On the sand-hills near Fort San Antonio. A stunted form, less pubescent than usual. Flowers buff.

Distribution. Yucatan, Guiana; Brazil, from Ceara to Rio de Janeiro; Peru; Jamaica and Senegal.

S. paniculata, *L. Sp. Pl.* p. 962; *DC. Prodr.* i. p. 465; *Cav. Diss.* i. p. 16.—S. atrosanguinea, *Jacq. Ic. Rar.* i. t. 136.

A common shrubby plant, with dark purple flowers. In many of the bushy places, on the main island, especially Chaloupe Bay, the Peak, and the Sapate.

Distribution. Jamaica; Peru; Ecuador; and Brazil, where it is common from Pernambuco? to Rio de Janeiro.

S. spinosa, *L. Sp. Pl.* p. 960; *DC. Prodr.* i. p. 460.

Frequent on the slopes of Chaloupe Bay, near the Fort.

Distribution. Cosmopolitan.

S. glomerata, *Cav. Diss.* i. p. 18, t. 2. f. 6.—S. carpinifolia, *DC. Prodr.* i. p. 460.

Along the main roads through the central district. Common.

Distribution. Widely distributed.

STERCULIACEÆ.

WALTHERIA AMERICANA, *L. Sp. Pl.* ed. 1, p. 673 ; *H. B. K. Nov. Gen. et Sp.* v. p. 333 ; *Schum. in Mart. Fl. Bras.* fasc. 96, p. 64, t. xii. fig. 1.

Common among thickets, Chaloupe Bay, and in the central district. Also at the base of the Peak.

Distribution. Whole of the tropical world. Very common in Brazil.

STERCULIA FŒTIDA, *L.*

There were one or two fine trees of this plant in the gardens in the village.

GERANIACEÆ.

OXALIS NORONHÆ, *Hook. Ic. Pl.* xiv. p. 21, t. 1226 ; *Hemsl. Bot. 'Challenger,' Exped.* pt. ii. *Atlant. Isl.* p. 14.

This plant is common on nearly all the larger islands, Ilha dos Ratos, Sella Giueta, and the main island, wherever it can find sufficient protection from the wind. It was originally described from imperfect material collected by Darwin and Moseley ; and is, as far as at present known, peculiar to this group of islands. I examined the plants with some care, and am therefore able to add some further notes concerning it. It is a shrub of from 1 to 6 feet in height, attaining its greatest dimensions in the Sapate woods, where it grows freely intermingled with *Jacquinia armillaris, Palicourea,* and other shrubs. The stem is never more than 2 inches thick, and covered with a smooth brown bark. The branches slender, rather stiff and erect. The leaves are very slightly sensitive, light dull green, and, like almost the whole plant, pleasantly acid. The flowers are large and bright yellow, opening in the early morning and closing up as the sun becomes hot, so that the flower looks again like a bud. There are two forms of the flower borne on different bushes, differing in the length of the styles. The commonest is the brevistyled form. In this the inner whorl of stamens is long enough to reach to the mouth of the corolla, while the outer row, which are thicker at the base, reach only about halfway. The bright green stigmas project between the upper stamens at a distance of about halfway between the anthers of the upper and lower whorl. In the long-styled form the stigmas are raised to the level of the corolla-mouth, while the long whorl of stamens is considerably shorter. At the base of the outer stamens glands secrete nectar, which is sought by a small black bee (*Andrena*, sp.), which is no doubt the fertilizer of the plant; for I never saw any other insect of sufficient size at the flowers.

The capsule is explosive.

OXALIS SYLVICOLA, n. sp. (Plate II. figs. 3, 4.)

Herba annua, erecta gracilis semipedalis raro ramosa, radice fibrosa. Caulis tenuis pubescens. Folia trifoliata; petioli gracillimi pubescentes ferme unciam longi, erecto-patuli; foliola late ovata obtusa brevissime petiolulata, $\frac{1}{2}$ unciam longa, $\frac{3}{8}$ unciam lata, parce pubescentia præsertim ad bases; stipulæ nullæ. Flores parvi pulchre flavi, 3–4 in apice pedunculi tenuis pubescentis $1\frac{1}{2}$-uncialis erecti, demum fructu maturante deflexi. Bracteæ minutæ setaceæ. Pedicelli longiores, $\frac{1}{8}$-unciales. Calyx pubescens, sepala 5 lanceolata acuminata angusta pubescentia vix $\frac{1}{6}$-unciales. Petala obovata unguiculata obtusa vel flava $\frac{5}{16}$ unciam longa. Stamina 10, interiora longiora, filamentis pubescentibus, exteriora brevia glabra; antheræ ovoideæ. Pistilla 5, quam stamina interiora breviora, glabra; styli breves, excurvi; stigmata capitata. Capsula quam sepala brevior pentagona. Semina pauca in loculo quoque, magna castanea oblonga transversim rugosa.

I have only seen as yet the short-styled form, which perhaps is the only form that exists, as is the case in *O. stricta* and some others; but these possess long-styled flowers only. The flower-peduncle is at first erect; but as the fruit ripens it becomes deflexed till it forms an acute angle with the stem. The capsule is very short and pentagonal, and the seeds are unusually large for the capsule.

This pretty little *Oxalis* was found only at one spot in the thickest part of the Sapate, at almost the furthest accessible point. There was only a small quantity of it growing among the bushes by a woodcutter's path.

SAPINDACEÆ.

SCHMIDELIA INSULANA, n. sp.

Frutex ramosus magnus foliosus cortice griseo verruculoso. Folia trifoliolata glabra polita læte viridia coriacea; petioli 1½-unciales foliola subsimilia ovata utrinque acuminata acuta, ferme integra, marginibus obscure sinuatis, lamina 3 uncias longa, 1½ unciam lata. Racemi simplices in axillis foliorum superiorum, nutantes, 1½ unciam longi, rhachide pubescenti, basi nudi. Flores laxi parvi virides, pedicellis brevibus pubescentibus. Bracteæ brevissimæ ovatæ obtusæ.— ♂ flores. Sepala 4, inæqualia lata ovata rotundata obtusa, margine pubescenti. Petala sepalis subæqualia, unguiculata, obcuneata, truncata, pubescentia. Stamina 8; filamenta quam sepala paullo longiora, basi incrassata pubescenti, antheræ ovoideæ. Discus rotundatus inæquilaterus, marginibus incrassatis involutis.

This large bushy shrub was very plentiful in the Sapate, especially in the more open spots. Its leaves are bright green and somewhat hard and polished. The flowers are small and green, borne on short racemes in the upper axils. The plant only commenced to flower shortly before we left the island; and although we sought carefully and took specimens from numerous bushes, we were unable to find any female flowers; nor is there the least trace of a pistil visible upon the disk in the male flowers.

CARDIOSPERMUM HALICACABUM, *L. Sp. Pl.* ed. 1, p. 366.

Common on the main island in the thickets, especially in the Sepate. Baskets are made of its stems. Flowers white, very sweet-scented.

Distribution. Most tropical countries.

AMPELIDEÆ.

VITIS VINIFERA, *L. Sp. Pl.* ed. 1, p. 202.

There are a few Vines cultivated here which produce good fruit, but not in large quantity.

V. SICYOIDES, *Baker in Mart. Fl. Bras.* xiv. 2. p. 202.—Cissus sicyoides, *L. Sp. Pl.* ed. 2, p. 170.

Abundant, climbing over the bushes in the main island in the more open parts. In a small wood of Burra in the centre of the island the stems of large size hung down from the trees like lianes. The flowers are cream-coloured; the berries black and sweet. In the thickets of the Sapate we frequently found the remarkable monstrous tufts looking like some parasitic plant on

the vine-stem, which were called by Presl *Spondylantha aphylla*
(Rel. Hænk. ii. 35. t. 53).

ANACARDIACEÆ.

SPONDIAS PURPUREA, *L. Sp. Pl.* ed. 2, p. 613; *Jacq. Amer.*
t. 131.

There are a number of trees of what seems to be this species
not only in the gardens, but also apparently wild in the Sapate,
perhaps planted there by birds. Many of the trees were quite
bare of leaves during our visit, and neither fruit nor flowers
were seen. It is known as " Caja."

MANGIFERA INDICA, *L. Sp. Pl.* ed. 1, p. 200.

There are a few trees of the Mango scattered about the
island.

ANACARDIUM OCCIDENTALE, *L. Sp. Pl.* ed. 1, p. 383; *Griseb.*
Fl. Brit. W. Ind. p. 176.

Is abundant in many spots in the central, cultivated districts,
growing often in the maize-fields, and also in the Sapate. It
does not appear to be indigenous here, as it is doubtless in
Pernambuco.

COMBRETACEÆ.

TERMINALIA CATAPPA, *L. Mant.* p. 519.

There are a few trees of this plant scattered over the island,
it having been introduced.

LAGUNCULARIA RACEMOSA, *Gaertn. f. Fruct.* iii. t. 217; *Eichl.*
in Mart. Fl. Bras. xiv. 2. p. 101.

The largest stream, the one at Suesta, which flows into the sea at
Tobacco Point, had a thick fringe of this Mangrove along its banks.
The trees were about 20 feet high, a good deal taller than they are
in the mangrove-swamps on the mainland.

Distribution. All the coasts of Tropical America and Western
Africa.

COMBRETUM, § TERMINALIOPSIS, n. sect.

Frutex ramosus *dioicus,* foliis oppositis coriaceis ovatis obtusis
exstipulatis, spicis axillaribus gracilibus, floribus minimis glo-
bosis viridibus pubescentibus sessilibus, sepalis connatis epigynis
intus ac extus pubescentibus, petalis nullis.

COMBRETUM RUPICOLUM, n. sp.

Frutex dioicus ramosus. Folia opposita ovata rotundata coriacea glabra obtusa, 3 uncias longa, 2 uncias lata, petiolo crasso ½-unciali. Stipulæ nullæ. Racemi 2–3-unciales in axillis foliorum basibus breviter nudis, rhachide pubescenti. Flores parvi copiosi virides sessiles pubescentes. Bracteæ minutæ lanceolatæ pubescentes, ovariis æquilongæ. Sepala 4, connata, apicibus rotundatis obtusis extus et intus pubescentia. Petala nulla. Stylus cylindricus integer sepala paullo superans, apice curvo; stigma parvum integrum. Ovarium quadratum pubescens, ovulum singulum erectum.

This shrub grows on the basaltic boulders of the East Hills, about 600 feet above sea-level. It had dark green opposite leaves and slender racemes of green flowers. It was only found in flower just previous to our departure; and we were unable to find any male flowers.

It is probable that it would constitute a new genus of Combretaceæ; but in the absence of male flowers and fruit, I think it unadvisable to found a new genus on our material. One or two Combretums show a tendency to become diœcious; but this is the only known truly diœcious species. The habit is somewhat that of a *Terminalia*; but the opposite leaves show it to be really nearer to *Combretum*.

LEGUMINOSÆ.

§ GENISTEÆ.

CROTALARIA STRIATA, *DC. Prodr.* ii. p. 131; *Benth. in Mart. Fl. Bras.* xv. 1. p. 26.

Common amongst the fodder-plants in the central district.

Distribution. Warm parts of the whole world. It is abundant in waste ground round Pernambuco.

§ GALEGEÆ.

INDIGOFERA ANIL, *L. Mant.* p. 272; *Vell. Fl. Flum.* vii. t. 20; *Benth. in Mart. Fl. Bras.* xv. 1. p. 41.

Not rare along the pathway through the Sapate, and by the edges of the maize-fields at Leao. Widely distributed all over the world, and probably introduced here by man.

TEPHROSIA CINEREA, var. LITTORALIS, *Benth. in Mart. Fl. Bras.* xv. 1. p. 48.—Vicia littoralis, *Jacq. Amer.* t. 124.

Common in the bushes at Chaloupe Bay and on the sand-hills
at Fort San Antonio. Also common in Rat Island. The flowers
are usually pink; but white-flowered specimens were found on
Rat Island. Like most of the genus, it kills fish when put into
the water in which they live. Bunches of the plant with long
roots were bruised with a club and stirred in a rock-pool. Pre-
sently the fish concealed in the coral clefts began to dart in and
out, and soon coming out entirely, went into convulsions, and
finally died. We obtained many species in this way, which would
otherwise have been difficult to obtain.

Distribution. All warm parts of South America from Mexico
southwards.

SESBANIA ÆGYPTIACA, *Pers.*

There were two or three trees of this beautiful plant loaded
with flowers in the convicts' gardens.

§ HEDYSAREÆ.

ÆSCHYNOMENE HISPIDULA, *H. B. K. Nov. Gen. et Sp.* vi. p. 531;
Benth. in Mart. Fl. Bras. xv. 1. p. 59.

This shrubby plant was very common in the central district of
the main island, and on the north side of Rat Island were dense
beds of it covering the slopes towards the sea to the exclusion of
other plants. The flowers are yellow.

Distribution. Central America to Minas Geraes and Lima.

ZORNIA DIPHYLLA, *Pers. Syn. Pl.* ii. p. 318, var. RETICULATA
GLABRA, *Benth. in Mart. Fl. Bras.* xv. 1. p. 81.

Common on the turf of the eastern promontory beyond Fort San
Antonio; also at Tobacco Point; and a slightly large-leaved form
among the stones on the road above the Residency garden.

Var. ELATIOR, *Benth. l. c.*

Among long grass in Chaloupe Bay.

Distribution. A common plant of world-wide distribution, very
abundant in Brazil. It is very variable; but all the specimens
collected in Fernando Noronha belong to the glabrous group.

DESMODIUM TRIFLORUM, *DC. Prodr.* ii. p. 334; *Benth. in
Mart. Fl. Bras.* xv. 1. p. 95, t. xxvi.—Sagotia triflora, *Duchass. &
Walp. in Linnæa,* xxiii. p. 738.—Nicolsonia reptans, *Meissn. in
Linnæa,* xxi. p. 260.

Turfy spots on the promontory.between Chaloupe Bay and San Antonio Bay; on the top of the cliffs opposite the Frade; and very abundant and tall in Leao Bay, behind the Fort.

The minute flowers are of intense deep blue, rarely white.

Distribution. East Indies, and perhaps introduced thence into South America.

DESMODIUM (§ NICOLSONIA) BARBATUM, *Benth. in Miq. Pl. Jungh.* i. p. 224, *et in Mart. Fl. Bras.* xv. 1. p. 95.—Hedysarum barbatum, *L. Sp. Pl.* p. 1055.

On the sides of Morro branco, and also on the eastern slope of Look-out Hill.

In both of these localities the soil was phonolite altered by contact with basalt ; and it was never found except on this rock. Flowers brilliant blue.

Distribution. East Indies ; also all parts of tropical America.

D. SPIRALE, *DC. Prodr.* ii. p. 332.

Very common among thickets in Chaloupe Bay and the entrance to the Sapate.   Also plentiful on Rat Island.   Flowers white or yellow.   Obtained also by Moseley.

D. INCANUM, *DC. Prodr.* ii. p. 332 ; *Benth. in Mart. Fl. Bras.* xv. 1. p. 98.

On the top of the cliff between Chaloupe Bay and San Antonio Bay.   Also common along the path through the Sapate.   Flowers rose-pink.

Distribution. All tropical countries.

§ VICIEÆ.

ABRUS PRECATORIUS, *Linn. Syst.* ed. xii. p. 472.

Very common on the Sapate, and also on the cliffs between Chaloupe Bay and San Antonio Bay.

Distribution. Throughout the tropics generally.

§ PHASEOLEÆ.

ERYTHRINA AURANTIACA, n. sp.   (Plate I.)

Arbor 20–30-pedalis, diametro 9–12 uncias.   Cortex lævis, atro-brunneus.   Coma patula ramosa.   Rami juvenes atri pulverulento-tomentosi, spinis brevibus conicis atris nonnunquam bifidis tecti.   Folia trifoliolata, glauca, vetusta glabra, ad 7 uncias longa, petiolis spinosis, foliolo terminali ovato subtriangulari obtuso, $3\frac{1}{2}$ uncias lato, 3 uncias longo, lateralibus inæquilateris ovato triangularibus, petiolulis $\frac{1}{4}$-uncialibus, folia juvenilia minora pulverulenti-tomentosa. Flores in racemis subterminalibus brevibus speciosi iis *E. glaucæ* subsimiles, minores. Calyx bilabiata pulverulenta, $\frac{3}{4}$ unciam longa, labio inferiore apice bifido quam superior longiore. Vexillum unguiculatum ovatum obtusum patulum, $1\frac{1}{2}$ uncias longum, $1\frac{1}{2}$ uncias latum, aurantiacum, venis viridibus. Alæ breves, auriculiformes, $\frac{1}{2}$-unciales, virides, rubro-marginatæ. Carina brevis viridis alis subsimilis subæqualis. Stamina 10, basi longiusculi connati kermosina, biuncialia. Pistillum andrœcio subæquale. Ovarium pubescens. Stylus gracilis roseus. Stigma rotundata parva viridia. Legumen 3–4-uncialis, apice longe acuminata 1–2-spermum. Semen $\frac{1}{2}$-uncialis oblongum, dorso carinati, atrum politum, circa hilum coccineum.

Main island, scattered bushes near the village and in the Sapate. One full-grown tree in the cocoa-nut plantation at Sueste. It also occurs at Sella Giueta.

This very interesting tree is called "Mulungu" by the inhabitants of the island, which name is also applied to *Erythrina Mulungu*, Mart., quite a different species, of which we saw a single fine-tree at Iguarassa near Pernambuco. It is scrupulously cut down by the convict woodcutters under the direction of the Governor, as it is stated that if rafts are made of it they become water-logged and sink in open sea in three days. This is no doubt the tree called *Erythrina exaltata* by Webster in the Voyage of the 'Chanticleer' as cited on p. 18, which, he says, is the largest tree in the island; and it is also, I believe, the plant intended by Moseley, Linn. Journ. Soc. Bot. xiv. p. 360. "I saw several specimens of a tree with rounded leaves of a bluish green and stout thorns: it had a Euphorbiaceous look. . . . One of the trees was about 20 feet high and 9 inches in diameter of trunk." Now, I believe, there is only one tree of it left of any size in the islands, and that is in the cocoa-nut grove at Sueste; and the young plants scattered over the wooded districts and thickets in the main island, and also Sella Giueta, do not show any signs of bearing flowers, as they are too young. The tree which Moseley saw near the village is cut down, only a bush remaining. It so seldom flowers, that I could not at first elicit from the convicts what the colour of the flowers was; and under the present regime it will, I fear, soon be extinct. The tree at Sueste is about 30 feet high, with a spreading head of branches, which, as it was in flower at our visit, bore then only a few leaves on the extremity of the branches. The bark in younger plants is dark green, and covered with strong blackthorn ; but in the lower part of the older tree it was brown and bare of the thorns. It is quite smooth, and not cut up into cracks, as is the case in *Erythrina Mulungu*. The full-sized leaves are glabrous and of a greyish green, the younger ones covered with a mealy pubescence which becomes rufous. The flowers are borne on short racemes, of about a dozen, on the ends of the branches. They somewhat resemble at first sight those of *E. glauca*, Willd., but are a good deal smaller. The standard is broad and reflexed, of a dull orange-colour, with greenish veins. The alæ and keel are polished green with red edges, and the bright crimson androecium contrasts beautifully with the orange standard. The buds are covered with a reddish tomentum like the young branches. We were unable to obtain fruits at the time of our visit ; but by the kindness of the Director of the island we received a good series of both fruits and seeds. The pods contain one or two seeds, entirely black, except for a red band round the hilum ; they are oblong, smooth, and polished. With the specimens came notes, saying that "it is said that three or four unbarked Mulungu-seeds, being ground and mixed with food, will kill any dog or cat that eats it ; and consequently these seeds are never found eaten by mice. It is propagated by cuttings. And a warm infusion of the inner part of the bark is used in toothache."

ERYTHRINA, sp.

Among the fruits and seeds sent after our return to England by the Director of the island were fruits and seeds of another species of *Erythrina* from Leao, with a note that this species very closely resembled the preceding in foliage and habit, but that the seeds were different. The pod is rather longer and broader, and less abruptly dilated where the seeds occur. The seeds are a little longer and more pointed, *i. e.* less oblong with no distinct keel, and entirely red except the hilum. This is the rosy Mulungu, "Mulungu vermelho" of the inhabitants. The material received is insufficient for determination as to species.

Mucuna urens, *DC. Prodr.* ii. p. 405 ; *Benth. in Mart. Fl. Bras.* xv. 1. p. 169, t. xlvi. 1.

We obtained two seeds of this plant among seaweed &c. drifted on the shores of Suéste Bay ; but the species has not established itself here yet. It is very common on the mainland, and was well known to the convicts, who all agreed that it did not belong to the island. This is well known as a drift-seed (Hemsley, ' Chall.' Rep. Bot., Juan. Fernand. &c. p. 299).

Canavalia obtusifolia, *DC. Prodr.* ii. p. 404 ; *Benth. in Mart. Fl. Bras.* xv. 1. p. 178, t. xlviii.—Dolichos obtusifolius, *Lam. Encyc.* ii. p. 295.

Plentiful on Rat Island on both sides, and on the main island at the base of the Peak and at Tobacco Point and beyond Morro branco. It grows only over the fallen boulders of basalt a little way above high-water mark, mingled with *Philoxerus vermicularis.* It is a large plant with showy crimson flowers. Like the preceding, this is a plant of wide distribution, the leaves of which are carried about by ocean-currents.

Phaseolus lunatus, *L. Sp. Pl.* i. p. 1016 ; *Benth. in Mart. Fl. Bras.* xv. 1. p. 181.—P. bipunctatus, *Jacq. Hort. Vindob.* t. 100.

Common, and apparently cultivated in the island. It occurred in many parts of the central district, and also at the summits of some of the more uncultivated hills, such as Morro branco.

Distribution. All warmer parts of the world.

P. peduncularis, *H. B. K. Nov. Gen. et Sp.* vi. p. 447 ; *Benth. in Mart. Fl. Bras.* xv. 1. p. 184.

One of the most abundant plants in the main island, and equally common on St. Michael's Mount and on Rat Island, covering extensive tracts of country.

Distribution. Central America and North Brazil.

§ Bauhinieæ.

Bauhinia forficata, *Link.*

" Mororo." One or two trees occurred in the village, introduced as timber-plants. The wood is said to be so hard that only the best tools will cut it ; and it is thus used as a test for axes and scythes.

SWARTZIA PINNATA, *Willd.*?

Webster, in his account of the island, mentions *Swartzia pennata* as an ornamental plant, and sweet-scented in the evening. We saw nothing answering to his account; and it was possibly the true *S. pinnata*, Willd., introduced in cultivation.

CAJANUS INDICUS, *Spreng. Syst.* iii. p. 248; *Lindl. Bot. Reg.* (1845), t. 31.

Several large plants near the village on the west side and in the sugar-cane fields in the central district. It is used as medicine, and was no doubt introduced intentionally.

RHYNCHOSIA MINIMA, *DC. Prodr.* ii. p. 385; *Benth. in Mart. Fl. Bras.* xv. 1. p. 204.

Common on the turf at the extreme eastern promontory at San Antonio Bay, and also at Tobacco Point.

Distribution. All warm countries.

§ CÆSALPINIEÆ.

CASSIA TORA, *L. Sp. Pl.* p. 538; *Benth. in Mart. Fl. Bras.* xv. 2. p. 115.

Very common in the central district, forming a good fodder for the animals; and perhaps introduced for that purpose.

Distribution. All warm countries.

C. OCCIDENTALIS, *L. Sp. Pl.* p. 539; *Benth. in Mart. Fl. Bras.* xv. 2. p. 113.

Common in the central district, and growing with the last species. The seeds are used to make a kind of coffee with some medicinal properties. The pods are collected and sold in the market tied up in small bundles for this purpose.

Distribution. All warm countries.

§ MIMOSEÆ.

ALBIZZIA LEBBEK, *Benth. in Hook. Journ. Bot.* iii. p. 87.

One or two trees occur in the gardens in the village. It is often grown in Pernambuco. It is called " Tamanqueira " because its wood is used to make wooden shoes of.

ACACIA FARNESIANA, *Willd. Sp. Pl.* iv. p. 1083.

There are numerous bushes of this plant on the shore below the village near the hospital, and also close to Fort San Antonio, and in thickets in the central district and at Sambaquichaba. It is known as " Coronha Christi " and " Espongeira ;" and the pods are used in making ink with the aid of iron; a gum is also extracted by boiling. It is a plant of world-wide distribution, and probably intentionally planted.

MYRTACEÆ.

PSIDIUM GUYAVA, *Raddi,* and JAMBOSA VULGARIS, *DC.*
Both are in cultivation here, the former fruiting abundantly.

LYTHRARIEÆ.

AMMANNIA LATIFOLIA, *L. Sp. Pl.* i. p. 174; *Koehne in Mart. Fl. Bras.* xiii. 2. p. 206.
Plentiful in one swampy spot in the central district, growing with *Paspalum brizoides, Jussieua,* and other marsh-plants.
Distribution. All warm parts of America.

ONAGRARIEÆ.

JUSSIEUA LINIFOLIA, *Vahl, Eclog.* ii. p. 31 ; *DC. Prodr.* iii. p. 55 ; *Micheli in Mart. Fl. Bras.* iii. p. 163, t. xxxiii.—J. acuminata, *Sw. Fl. Ind. Occ.* p. 245.
Very common in the central districts, on the damp clayey soil. The flowers are bright yellow. The leaves seem a little broader than usual. Obtained also by Moseley.
Distribution. Cosmopolitan. In Brazil it seems to be especially common in the north-eastern district, Para, Amazonas, Goyaz, and Pernambuco.

PAPAYACEÆ.

CARICA PAPAYA, *L.*
The Papaw is very largely cultivated, and its seeds apparently being carried about by birds, it is often to be seen in places where it appears to be quite wild. It is a very conspicuous feature in the scenery. The fruit is pyriform, and hangs down on the ends of the long peduncles ; it is remarkably good and cheap. The rats are very fond of it, and often climb trees to eat it. Near the village are several male trees which bear monœcious flowers, and can often be seen in fruit.

CUCURBITACEÆ.

LUFFA CYLINDRICA, *M. Roem. Syn.* fasc. 2, p. 63 ; *Cogn. in DC. Monogr.* iii. p. 456.

This is constantly cultivated in the gardens, being allowed to grow on the hedges and walls. The fibrous network is used as wadding for guns under the name of "Buchu."

LUFFA PURGANS, *Mart. Syst. Mat. Med. Bras.* p. 81 ; *Naud. Ann. Sc. Nat.* sér. 4 xii. p. 125.—L. operculata, *Cogn. in Mart. Fl. Bras.* vi. 6. p. 11.

"Cabacinha." Used here as a purge and emetic ; but a dangerous drug, as persons have been killed by the use of it.

[MOMORDICA CHARANTIA, *L. Sp. Pl.* ed. i. 1009.

Rat Island ; see p. 3.

Distribution. All tropical countries.]

CUCUMIS MELO, *L. Sp.* ed. 1, p. 101 ; *Cogn. in DC. Monogr.* iii. p. 482.

The Melon is very carefully cultivated, and produces exceedingly good fruit, far better than that of the mainland.

CITRULLUS VULGARIS, *Schrad. in Linnæa*, xii. p. 412 ; *Cogn. l. c.* p. 588.

This is also very productive here. I found a plant in which all the flowers were converted into leaf-buds.

CUCURBITA PEPO, *L. Sp. Pl.* ed. 1, p. 1010 ; *Cogn. in DC. Monogr.* iii. p. 545.

This is the most commonly cultivated of the three last-named. All these Melons, except *Luffa*, grow on all the open spaces, climbing up to the top of the hills, and are often to be found mixed up with endemic and other indigenous plants ; so that it is at first difficult to see which is cultivated and which wild. None of these occur on any of the other islands except Rat Island, which was famed for its Melons in Webster's time. The rats destroy a great quantity of them.

CAYAPONIA TAJUJA, *Cogn. in DC. Monogr.* iii. p. 772.—Bryonia Tajuja, *Vell. Fl. Flum.* x. t. 89.—Trianosperma Tajuja, *Mart. Syst. Med. Bras.* p. 80 ; *Naud. in Ann. Sc. Nat.* sér. 4, xvi. p. 192.

This was abundant on the Peak, growing mixed with *C. racemosa, Momordica*. The flowers are greenish, fruit orange. It is called "Tajuja" by the natives.

Distribution. Goyaz to Rio Grande do Sul.

CAYAPONIA RACEMOSA, *Cogn. in DC. Monogr.* iii. p. 768.—
Bryonia racemosa, *Sw. Prodr.* p. 116.—Cionandra racemosa,
Griseb. Fl. Brit. W. Ind. p. 286.—Trianosperma racemosa,
Griseb. Cat. Pl. Cub. p. 112.

Common on the Peak and at Tangle Rock, and in the Sapate.

Distribution. A native of Mexico, the West Indies, and
Guiana.

CERATOSANTHES ANGUSTILOBA, n. sp.

Dioica? caulis validulis, cirrhi longi graciles simplices. Folia
trifoliolata, foliolis lateralibus inæqualiter bilobis, lobi angusti
lineares lanceolatæ obtusi marginibus ciliatis; foliolo mediano
integro anguste lineari-lanceolato acuto, 3 uncias longo, ¼ unciam
lato. Pedunculi glabri validuli, 5-unciales. Racemi breves,
masculi pauciflori, compacti. Flores virescentes parvi, masculi,
tubus gracilis ⅜ unciam longus. Sepala glabra ovata subacuta
viridia; petala basi oblonga angusta, laciniis angustis lineari-
bus, pubescentia, ¼ unciam longa. Antheræ oblongæ breves.
Flores feminei et fructus non visi.

On the Peak, with other species.

This plant is easily distinguished from the rest of the genus
by its narrow lobed leaves, with the lateral lobes again bilobed.
The flowers also are very small in comparison at least with
those of *C. trifoliolata*, the only species really near to it. It had a
very unpleasant smell when bruised.

C. CUNEATA, n. sp.

Dioica, glabra, caules graciles striati, cirrhi simplices nec
valde longi. Folia triloba glabra, foliola obcuneata apice dentata
irregulariter, dentibus subacutis, 3 uncias longa, 1½ uncias lata
quo latissima. Pedunculi validuli 3-unciales, apice racemosi.
Racemus circiter 12-florus. Flores albescentes, perianthio in
utroque sexu simili. Tubus ½ unciam longus. Sepala brevia, late
ovata obtusa carnosula glabra. Petala 5 longa basi lata, lamina
angusta obtusa longe bifida, pubescentia, extus septem nervia.
Stamina oblonga brevia compacta, cohærentia. Flores feminei.
Stylus longus ferme ½-uncialis planus latus; stigmata 2, brevia
acuta; staminodia duo parva ovata perianthio adnata, paullo
supra stigmata. Bacca non visa.

On the Peak, with other species of Cucurbitaceæ.

This species is allied closely to *C. Hilariana*.

CERATOSANTHES RUPICOLA, n. sp.

Dioica, glabra, caules graciles striati, cirrhi longi simplices. Folia triloba, marginibus ciliatis, lobi profunde fissi nec omnino liberi, exteriores obliqui ovati angulati integri minute mucronati, 2 uncias longi, 1 unciam lati, petiolis ¾-uncialibus. Pedunculi 6-unciales, glabri, racemi breves 1½-unciales. Flores masculi . . . albescentes, 1¼ unciam longi pedicellis inclusis; sepala 5 ovata brevia submucronata crassiuscula, extus pubescentia; petala basi oblonga, laciniis longis angustis obtusis, ½-uncialia, extus pubescentia intus ad basin, costis elevatis 5. Stamina tubo adnata. Antheræ breves crassi, ⅛ unciam longæ. Flores feminei et fructus non visi.

This plant only occurred scrambling over the basalt rocks of the East hills. It is allied to *C. trifoliolata*, Cogn., but distinct in its less deeply cut leaves and entire leaflets.

FICOIDEÆ.

SESUVIUM DISTYLUM, n. sp.

Herba prostrata læte virens, carnosa, ramis 6–12-uncialibus. Folia cuneata spathulata obtusa, exstipulata, basi scariosa vaginante, unciam longa, ¼ unciam lata. Flores albi singuli breviter pedicellati, ¼ unciam in diametro. Perianthium basi connatum, tubo brevi, segmenta tria exteriora lanceolata acuta, submucronata apice cucullata, interiora duo latiora ovata mucronata. Stamina plura quam perianthium breviora, filamenta complanata tenuia; antheræ ovoideæ. Ovarium breve conicum obtusum, apice depresso. Styli duo graciles quam perianthium multo breviores. Semina plura nigra reniformia.

Rat Island, growing only within the spray of the blowhole at the south-west corner; abundant there, but seen nowhere else. The soil at this spot is reef-rock covered with guano in parts. The plant grows in the holes formed by the weathering of the rock, and makes bright green patches visible for some distance.

The species is very near *S. portulacastrum*, L.; but I believe it to be quite distinct in the colour of the flowers, which are not even tinted with pink or purple, and in the number of styles, never less than three in *S. portulacastrum*; but in this species there are but two, and these much shorter than in the latter plant.

CACTACEÆ.

CEREUS INSULARIS, *Hemsl. Bot. 'Chall.' Exp., Atlantic Isles*, p. 16, t. xiv.

This was described and figured from material brought by the 'Challenger' expedition; but as the description is incomplete in some respects and inaccurate in others, owing to deficiency of material, I append an emended description.

Planta valida, ramis 6–12-pedalibus, erectis vel pendulis, apicibus recurvis, 1–1½ uncias diametro, costis obtusis sæpissime 7, ¼ unciam altis continuis, areolæ pulvinatæ, ¼ unciam dissita lanugine albo parvo; spinæ radiatæ ochreæ 12–15, inæquales pungentes, longiores, unciam longæ. Flores nocturni, 5–6 uncias longi, 2 uncias in diametro, erecti, amœnissime odori, ochroleuci. Calycis tubus 3-uncialis, vix ¼ unciam in diametro, viridis, squamis paucis rufescentibus lanceolatis acuminatis. Sepala lanceolata acuta pauciseriata, rubro-viridia, longiora uncialia. Petala ochroleuca ferme alba tenuia late lanceolata obtusa, apicibus minute fimbriatis, quam sepala paullo breviora. Stamina erecta copiosa, quam petala breviora; interiora multi breviora, filamenta alba apice attenuata; antheræ oblongæ, ⅛-unciales, flavi. Stylus validulus 5-uncialis. Stigmata ¼-unciam longa, 7–11, subacuta, viridia. Ovarium viride ¼-unciale multiovulata. Fructus magnus 4 uncias longus, 3 uncias latus, oblongus obtusus, kermesinus, albo-pruinosus. Semina parva nigra.

This is very common on Rat Island, Booby Isle, San José, Sella Gineta, as well as on the main island, and is one of the most conspicuous plants. It frequently hangs down over the cliffs, quite covering them; but also forms thickets in open dry spots, and also grows even in the dense thickets of the Sapate. Here there was a woodland form, which differed in some respects from the common type, and may be the *Cactus quadrangularis* of Webster. It has more slender dark green stems less than an inch in diameter, with five ridges only and very short weak thorns, the largest of which were only half an inch long. No signs of fruit or flower were found on this form; but I have little doubt that it is a wood form of the common species. On two occasions fasciated branches were found in the dense thicket at the summit of Tangle Rock.

The flower opens widely only at about 10 o'clock at night, at which time it is very sweetly scented. It is erect, of a pale cream colour within, and not yellow as described in the 'Challenger' Report. The sepals, however, are reddish green, and the calyx-tube and ovary bright green. The flower closes again before dawn, and apparently does not reopen during the following night. The stamens are very numerous, white, with yellow anthers, and erect, scattered over the corolla-tube, the inner ones much shorter than the outer; all are a little shorter than the petals. The styles lie outside the staminal ring, even when the flower is erect. The stigmas are short and thick, variable in number, green or greenish yellow, sometimes, but rarely, branched. They never appeared hooked as figured in the 'Challenger' Report. The fruit is large when fully developed, of a dark rose-pink, with a bloom on the outside. The placentæ are white and sweetish in taste. The seeds numerous and black.

The plant seems to be quite irregular in its periods of fruiting and flowering, as it was possible to obtain flowers, fruit, and young buds all at once on the same spot. The common name for it is "Chique chique."

GAMOPETALÆ.

RUBIACEÆ.

SPERMACOCE PARVIFLORA, *Hemsl.*, *Biolog. Centr.-Amer.*, *Botany*, ii. p. 59.—Borreria parviflora, *W. Mey. Fl. Essequib.* p. 81, t. i.

Common on the Peak and in the open spaces at Tangle Rock and elsewhere, covering large tracts with a tangled mass of stems, often knee-deep. The flowers are white.

Distribution. All Tropical America.

GUETTARDA LEAI, n. sp.

Frutex humilis, ramis brevibus, undique foliosus. Folia ovalia elliptica obtusa, pagina superiore glabra, inferiore glauco pubescens, costis prominulis, lamina ferme 4 uncias longa, $2\frac{1}{4}$ lata, petiolus puberulus $1\frac{1}{2}$ uncias longus. Pedunculi folia vix superantes, puberuli. Flores circa 10 parvi albi pubescentes. Bracteæ lanceolatæ acuminatæ, calyces vix æquantes. Calyx pubescens, truncatus, unidentatus, $\frac{1}{8}$ unciam longus. Corolla brevis, $\frac{1}{2}$ unciam longa, gracilis sericeo-pubescens alba. Lobi rotundati breves, interne glabri. Stamina 5 .... antheræ dorsifixæ multo longiores quam latæ, os tubi non superantes, ferme omnino sessiles in corolla. Stylus stamina non superans pubescens, basi incrassata. Stigma rotundatum clavatum. Drupa parva globosa testaceopuberula, calyce coronata.

This shrub is distinguished from all the other Brazilian species by its small flowers and large leaves. The foliage resembles that of *Guettarda viburnoides*; but the flowers are small, straight, and regular, like those of *G. angelica*, which also it resembles in fruit. It is a low leafy shrub, only growing in the denser parts of the wooded Cape Placellière. Very few plants were seen, and some of these not in flower or fruit. The flowers are white, about a dozen together in the cyme; the stamens as nearly as possible sessile on the corolla-tube, and the stigma reaching to the level of the top of the anthers.

I have much pleasure in associating with it the name of the Rev. T. S. Lea, who first found it.

PALICOUREA INSULARIS, n. sp.

Frutex gracilis 10–12-pedalis, undique foliosus. Folia coriacea, lanceolata, subobtusa densa, lamina 3 uncias longa, 1½ uncias lata; petiolus ¼-uncialis. Stipulæ breves amplexicaules unidentatæ. Inflorescentia axillaris. Panicula pauciflora patula vix triuncialis. Flores ⅜-unciales, albi, pedicellis ½-uncialibus patulis. Calyx poculiformis dentibus 5 parvis, subobtusis, viridis. Corolla 5-partita, laciniis loratis obtusis recurvis. Stamina quam petala breviora 5, libera ad basin; antheræ lineares acuminatæ non appendiculatæ. Stylus petala superans. Stigma clavatum obtusum viride integrum. Drupa viridis, 5-angulata, 5 loculis.

Main island, in the Sapate woods, only a few bushes at one spot.

COMPOSITÆ.

BLAINVILLEA RHOMBOIDEA, *Cass. in Dict. Sc. Nat.* xxix. p. 494; *DC. Prodr.* v. p. 492; *Oliver, Fl. Trop. Afr.* iii. p. 375.

Central district in damp spots. Also a garden-weed.

Distribution. Brazil, from Pernambuco to Rio de Janeiro; also Tropical Africa.

ECLIPTA ERECTA, *L. Mant.* p. 286; *DC. Prodr.* v. p. 490.—
E. alba, *Hassk. Pl. Jav. Rar.* p. 528.

Base of Peak; also below Tangle Rock in damp places.

Distribution. All warm and tropical countries.

ASPILIA RAMAGII, n. sp.

Suffrutex humilis, ramosus, pubescens, subpedalis. Folia opposita, ovata acuta, lamina in petiolo decurrente, margine minute
serrato, scabra, venis et petiolo hispidis, venis dorso prominulis
tribus. Lamina ferme 3 uncias longa, 1½ uncias lata. Petiolus
¾ unciam longa. Capitula in pedunculis 1–3-uncialibus hispidis
vix unciam in diametro. Involucrum 2-seriatum, serici externa
lanceolata hispida, interiora oblonga obtusa apice minute pubescenti, scariosa. Flores radii flavi, lamina lata oblonga apice
bifidi? 5 millim. lati, 15 longi, plurivenia; tubus brevis. Stylus
teres; stigmata gracilia subteretia. Flores tubulosi, plures,
flavi. Calycis lobi breves, obtusi. Corolla 5-partita, apicibus
acutis, marginibus incrassatis, dorso pubescentia; stamina libera,
filamentis gracilibus. Antheræ elongatæ atræ, lineares, con-
nectivo apiculato, corollam superantes. Stylus teres : stigmata
crassiuscula, apice abrupte acuminata.

East Hills.

I am pleased to associate this plant with the name of Mr. G.
A. Ramage, who found it on the summit of the Eastern Hills,
growing among the broken basaltic boulders. The leaves, when
crushed, were deliciously aromatic : but the scent disappeared on
drying. The plant is a small half-shrub, with rough scabrid
foliage. The blade of the leaf is decurrent on the petiole, so that
the three prominent veins on the back of the leaf do not arise
from the base of the lamina. The flowers are bright yellow.

AGERATUM CONYZOIDES, *L. Sp. Pl.* p. 1175; *DC. Prodr.* v.
p. 108 : *Baker in Mart. Fl. Bras.* vi. 2. p. 194.

Exceedingly common all over the central district; but not
found on any of the other islands. It is the commonest species
of Compositæ here, covering large tracts of ground.

Distribution. World-wide.

ACANTHOSPERMUM HISPIDUM, *DC. Prodr.* v. p. 522.

There were a number of bushes of this plant in and about the
village. It seems usually to occur in waste spots and sandy shores
in various parts of South America, including Brazil.

PLUMBAGINEÆ.

PLUMBAGO SCANDENS, *L. Sp. Pl.* ed. i. p. 215 ; *H. B. K. Nov. Gen. et Sp.* ii. p. 220 ; *J. A. Schmidt in Mart. Fl. Bras.* vi. p. 166, t. xlvi. fig. 2.—P. occidentalis, *Sweet, Hort. Brit.* ed. 3, p. 565.

There was a large patch or two on the sand-hills at San Antonio, and also among the bushes at the entrance to the Sapate near the path.

Distribution. South America, from Yucatan to Rio de Janeiro. It is common in the woods at Pernambuco.

MYRSINEÆ.

JACQUINIA ARMILLARIS, *Jacq. Amer.* p. 53, t. 39 ; *Linn. Sp. Pl.* ed. 2, p. 272 ; *Miq. in Mart. Fl. Bras.* x. p. 281.

One of the commonest bushes in the Sapate, attaining a height of 8 or 9 feet. Stunted bushes occurred also on Morro branco, and on the hillside at Tangle Bay. The fruit is a rather sweet cherry-red berry. The flowers are pale flesh-colour.

Distribution. West Indies and as far south as Bahia.

SAPOTACEÆ.

BUMELIA FRAGRANS, n. sp.

Arbor ramosus ad 20-pedalis, spinosus ; ramis junioribus ferrugineo-pubescentibus. Folia juniora cuneata spathulata obtusa, $\frac{3}{4}$ unciam longa, $\frac{3}{8}$ unciam lata, seniora oblonga obovata obtusa, unciam longa, $\frac{3}{4}$ unciam lata, omnia atro-viridia, lucida, dorso pubescentia. Flores circiter 20 vel pauciores in glomerulis pallide virides, amœnissime odores, pedicellis vix $\frac{1}{4}$-unciales argenteosericeis. Sepala 5, ovata sericea integra. Petala exteriora 5 late ovata obtusa glabra, marginibus minute crenulatis ; interiora 10 angustata lanceolata breviora, margine dentato. Stamina 5, petala superantia ; antheræ angustæ oblongæ, extrorsæ, filamentis crassiusculis ; staminodia 5, petalis interioribus majora, lanceolata tenuia marginibus denticulatis, connectiva crassa. Stylus acuminatus subacutus, basi crassa. Ovarium pilis circumcinctum. Drupa viridis oblonga, ferme $\frac{1}{2}$ unciam longa. Semen durissimum atrum lucidum.

This plant grows in the Sapate as a thorny compact shrub, or in more open spots as a large bushy tree with the habit of a blackthorn. It is very thorny, the spines about half an inch long. The leaves are dark green, somewhat shiny and pubescent on the back; on the younger bushes cuneate, and broader and more rounded on the bigger trees. The flowers are green and inconspicuous, but most deliciously scented, and, being very profuse, the whole tree is strongly perfumed when in flower. The flowers of this species and, as far as one can make out from dried specimens, those of the other two Brazilian species are proterogynous. The berries are oval and green, like very small unripe sloes. The plant is called " Quichaba " by the inhabitants, which name is referred by Miers, in his manuscript list of woods of Brazil preserved in the British Museum, to some species of Sapotaceous plant unknown to him, but which was probably the common *Bumelia obtusifolia.* It gives its name to one of the settlements of the island, viz. Sambaquichaba, *i. e.* Chan de Quichaba, the plain of the Quichaba; but we did not see any specimens there during our visit. The tree seems to prefer stony and even rocky ground on the exposed cliffs or in the thickets of the woods.

ACHRAS SAPOTA, *L.*
The Sapota is cultivated in the gardens, and fruits well.

ASCLEPIADEÆ.

GONOLOBUS MICRANTHUS, *Hemsl. 'Chall.' Report, Atlant. Isles,* p. 18, pl. xv.

The endemic plant is very common, clambering over the bushes in the more open spots, especially on the Burra shrubs. It occurs on the main island, Rat Island, and Sella Gineta.

The stems are covered with a thick corky bark. The flowers are small and green, with black spots at the base of the petals. The endemic Tyrant, *Elainea Ridleyana,* Sharpe, uses the pappus of the seeds to line its nest with.

LOGANIACEÆ.

SPIGELIA ANTHELMIA, *L. Sp. Pl.* ed. 1, p. 149; *Lam. Ill.* t. 107; *Prog. in Mart. Fl Bras.* vi. 1. p. 262.

In sandy spots, under the cocoa-nut palms, at Leao, and tolerably abundant in one spot. No doubt introduced. It is a common Brazilian weed.

APOCYNACEÆ.

RAUWOLFIA TERNIFOLIA, *Kunth, Syn. Fl. Æquin.* ii. p. 298; *Griseb. Fl. Brit. W. Ind.* p. 408.

Common on the main island in the open spots, especially in the central district. The flowers are pinkish white, the berries pink and of somewhat nauseous taste. The plant is called " Frutta di Sapo."

Distribution. Tropical South America.

VINCA ROSEA, *L. Sp. Pl.* p. 305.

This is grown in gardens, and has wandered a short way from them and half established itself in one or two spots near the village.

GENTIANEÆ.

SCHULTESIA STENOPHYLLA, *Mart. Nov. Gen. et Sp.* ii. p. 106, t. 182, *et Prog. in Mart. Fl. Bras.* vi. 1. p. 207.—Reichartia rosea, *Karsten, Fl. Columb.* i. p. 59, t. xxix.

Very common in the central district, growing with *Ageratum conyzoides*, L., and other weeds. It also occurred on Morro branco. The flowers here were of a lurid pinkish cream-colour, while those found on the mainland of Brazil were almost yellow. We never found them coloured as shown by Martius or Karsten.

Distribution. Panama, West Indies, Guiana, Brazil, and Sierra Leone.

BORAGINEÆ.

HELIOPHYTUM INDICUM, *DC. Prodr.* ix. p. 556.—Heliotropium indicum, *L. Sp. Pl.* ed. 1, p. 130; *Griseb. Fl. Brit. W. Ind.* p. 485.

Is very common in the central district. It is known as " Fedegozi," and the leaves are made into a kind of tea for chest-complaints.

Distribution. Very common all over the world.

CORDIA GLOBOSA, *H. B. K. Nov. Gen. et Sp.* iii. p. 76; *Griseb. Fl. Brit. W. Ind.* p. 481; *Browne, Hist. Jam. Pl.* t. 13. fig. 2.

A shrub in the more open parts of the Sapate. Among the bushes, with white flowers and orange berries.

Distribution. W. Indies, Mexico, and Panama.

Apparently not hitherto recorded from Brazil; but a specimen collected by Gardner appears to be this plant.

CONVOLVULACEÆ.

Ipomœa Tuba, *G. Don, Syst.* iv. 271 (1837); *Meissn. in Mart. Fl. Bras.* vii. p. 216.—I. grandiflora, *Lam. Ill.* i. 467 (1791).— Convolvulus Tuba, *Schlecht. in Linnæa*, 1831, p. 735.—Calonyction grandiflorum, *Choisy, Conv. Or.* p. 60; *DC. Prodr.* ix. p. 346.

Forms a large densely matted bed covering the rocky débris in Chaloupe Bay, where the stems attain a thickness of nearly 2 inches and a considerable length, exuding a copious white latex when cut. It is almost equally abundant on a slope at the base of the Peak; and occurred also on Platform Island. This is called " Salso da Praia " here.

Distribution. Antilles, Surinam, and Guiana; but it does not appear to be known from Brazil except from this locality. It was only found on the north side of the island, and probably the seeds were drifted on to the island from the north.

I. muricata, *Jacq. Hort. Schœnbr.* iii. p. 40, t. 323, *non H. B. K.*—Calonyction speciosum, var. muricatum, *DC. Prodr.* ix. p. 345.

On the extreme north of the Sapate among bushes, on Cape Placellière; also in the central district. Flowers pink; open in the evening.

Distribution. World-wide?

I. Quamoclit, *L. Sp. Pl.* p. 227.

Occurs in the gardens, where it is cultivated to a small extent under the name of " Prima vera."

I. Pes-Capræ, *Sweet, Hort. Sub. Lond.* ed. 2, p. 289; *Meissn. in Mart. Fl. Bras.* vii. p. 256.

This is very common in all the sandy bays on the main island; especially abundant on the sand-hills at San Antonio and at Sambaquichaba, Portuguese Bay, &c. Some of the stems attained a length of 30 feet.

Distribution. World-wide.

I. Batatas, *Lam. Encyc.* vi. p. 14; *Meissn. in Mart. Fl. Bras.* vii. p. 281.—Convolvulus Batatas, *L. Amœn. Ac.* vi. p. 121.

The Sweet Potato is cultivated in many parts of the main island
and has run half wild in many spots; but it is most abundant on
Rat Island at the place covered with guano. Here it is very
profuse, owing to the richness of the soil; and the main island is
supplied with the tubers from this spot.

IPOMŒA DIGITATA, *L. Sp. Pl.* p. 228; *Meissn. in Mart. Fl.
Bras.* vii. p. 278.—Batatas pauiculata, *Choisy, Conv. Or.* p. 54.
Tolerably common in spots in the central district, growing
among the Leguminosæ.
Distribution. Throughout the tropics.

I. PENTAPHYLLA, *Jacq. Coll. Bot.* ii. p. 297; *Ic. Pl. Rar.* ii.
p. 319.—Convolvulus pentaphylla, *L. Sp. Pl.* ii. p. 223; *Meissn.
in Mart. Fl. Bras.* vii. p. 287.
This Convolvulus was very common in the central district of the
main island, and also in Rat Island.
Distribution. Over the whole world.

JACQUEMONTIA EURICOLA, n. sp.
Herba prostrata vel suberecta, ramosa pubescentia glauca undique tecta. Folia cordata ovata acuta, 2 uncias longa, 1½ uncias
lata, petiolus uncialis. Flores congesti in capitula parva, pedunculis 2-uncialibus suberectis axillaribus et subterminalibus.
Bracteæ plures, lanceolatæ acutæ, circiter ¼-unciales. Sepala 5,
pubescentia valde inæqualia 2 ovata lanceolata acuminata ¼-un-
cialia, 3 minora ovata acuta. Corolla campanulata ¾ unciam
longa, pallide azurea, ½ unciam in diametro. Stamina 5, pistillum
haud superantia. Anthera ovalis, loculis haud disjunctis. Ovarium conicum obtusum. Stylus gracilis rectus; stigma bilobum,
lobis ovalibus.
Rat Island, southern side; main isle, on the southern side,
from near San Antonio to Tangle Bay, growing on open spaces
facing the sea.
A very pretty delicate lavender-blue Convolvulus, almost white
at times, with stem and leaves covered with a close grey
pubescence.

CUSCUTA AMERICANA, *L. Sp. Pl.* p. 180; *Jacq. Amer.* p. 24;
Meissn. in Mart. Fl. Bras. vii. p. 376, t. cxxvi. fig. 1.
Not rare on the main island; parasitic on *Cucumis Anguria*, L.,

in Chaloupe Bay, and on *Ipomœa Pes-Capræ* in Pirate's Creek;
also Rat Island.

Distribution. Common in Tropical America.

CUSCUTA GLOBOSA, n. sp.

Caules tenues longi rubri. Flores perparvi $\frac{1}{8}$ unciam longi,
virescenti-albi, papillosi, in glomerulis dense congesti unciam in
diametro. Pedicelli teretes ferme $\frac{1}{4}$-unciales. Bracteæ ovatæ
acutæ papillosæ. Sepala 4 carnosula ovata subacuta, erecta.
Petala 4 haud reflexa tenuiora haud multo longiora angustiora
oblonga, tubo brevissimo. Stamina 4, petala non superantia, filamenta basi incrassata. Antheræ rotundatæ brunnescentes. Staminodia? brevia digitata subspathulata. Ovarium globosum
apice depresso. Styli 2, inæquales graciles, ovarium multo longiores. Capsula globosa parva. Semina 2.

Main island, only parasitic upon Leguminosæ, *Æschynomene*,
Philoxerus, and *Amaranthus*, &c. Summit of Morro branco and
near Tangle Bay.

This species 1 at first thought might be a form of *C. decora*;
but on examining that species, I found that the flowers were considerably larger, the petals reflexed, and the ovary almost conical.
C. globosa is remarkable for its very small flowers in dense balls
clustered on the branches of the host, the stems, which are very
slender, soon disappearing. The petals, sepals, and bracts are
covered with little papillæ arranged in lines. The petals are not
recurved, but almost connivent, very deeply cut, so that there is
hardly any tube.

SOLANACEÆ.

CAPSICUM FRUTESCENS, *Willd. Sp. Pl.* i. p. 1050; *Sendtn.
in Mart. Fl. Bras.* x. p. 142.

The Capsicum, which is commonly cultivated, is very abundant
in a half-wild state in the Sapate and other bushy spots, the seeds
being apparently scattered about by the doves which devour it.

C. sp. ? is also cultivated in gardens, and fruits well.

LYCOPERSICUM ESCULENTUM, *Mill. Gard. Dict.* ed. viii. n. 2.

Is cultivated also to a considerable extent, and has also run half-
wild everywhere. The half-wild form is a small plant, with round
orange berries as large as a cherry; and it is stated that the large-

fruited forms cultivated here speedily revert to the small-fruited form.

DATURA STRAMONIUM, *L. Sp. Pl.* ed. 1, p. 179.

It is common in and round the village. It is called "Stramondi," and used in medicine. The flowers open a little after 6 o'clock, *i. e.* just after sunset. I never saw any insects at them, though we watched them. It fruits extensively.

D. FASTUOSUM, *L.*

Is cultivated in gardens as an ornamental plant.

NICOTIANA TABACUM, *L.*

A little tobacco is grown here, but of inferior quality.

PHYSALIS VISCIDA, n. sp.—P. hirsuta, var. ?, *Hemsl. Exped. 'Challenger,'* pt. ii. p. 19.

Herba suffruticosa, pedalis vel brevior, sæpe omnino glanduloso-pubescente, pilis simplicibus. Folia ovata subacuta dentata, 1–1½ uncias longa, ¾ unciam lata, petioli unciali. Flores parvi flavi immaculati nutantes, iis *P. minimæ* æquales, pedicellis gracilibus ¼ unciam. Sepala 5 lanceolata acuminata hispida sub anthesi non dilatata. Corolla pallide flava, ½ unciam longa, petala obtusa pubescentia præsertim intus campanulata. Stamina linearia flava angusta. Stylus gracilis, apice acuminato. Stigma capitatum. Bacca parva globosa viridis; calyx fructifer ovatus uncialis parce pubescens, lobis acuminatis angustis, pubescentioribus. Semina plana rotundata reniformia brunnea punctata.

Common in various parts of the main island in bushy spots. When growing in dry rocky or sandy spots it is more stunted and spreading and very viscid, covered all over with the glandular hairs. In the more bushy spots it is taller, slenderer, and more herbaceous, and less pubescent. It seems to be quite distinct from any form of *P. hirsuta*, Dunal, as it has neither violet anthers nor a spotted corolla. From *P. minima*, L., again it differs in the shape of the calyx in fruit, which has an ovate-acuminate outline, the free portion of the sepals being narrow, lanceolate acute, and very hairy. The interior of the corolla is very pubescent.

SOLANUM OLERACEUM, *Dunal, Syn.* p. 12; *Dunal in DC. Prodr.* xiii. p. 51.— S. nigrum, var., *Sendtn. in Mart. Fl. Bras.* x. p. 17.

In several spots in the Sapate, but always near the convicts' huts. The flowers are small and white. It certainly looks very unlike any form of the European *S. nigrum*.

Distribution. Brazil, common and widely spread.

SOLANUM PANICULATUM, *L. Sp. Pl.* ed. 1, p. 267; *Dunal in DC. Prodr.* xiii. p. 278; *Sendtn. in Mart. Fl. Bras.* x. p. 80.

About the village in waste places and at San José (Platform Island), abundant in the ruined fort. A loose little branched straggling shrub about six feet high, with violet or almost white flowers.

Distribution. Brazil; common round Pernambuco.

S. MAMMOSUM, var. CORNICULUM.—S. cornigerum, *André in Rev. Hort.* 1868, p. 33, *non Dunal* (S. corniculatum *in tab.*).

The true "Jurubeba." This plant occurs in waste places round the village. The fruit is used in liver complaints. The name "Jurubeba" is commonly applied to any *Solanum* of this group; but this, I was assured, was the correct plant. It is doubtless introduced here. This species seems not to be specifically distinct from *S. mammosum*, which, however, is not hitherto known from Brazil. The form of the fruit, which is rather larger than *S. mammosum*, and has an irregular number of processes at the base, seems to be the only distinguishing mark.

S. BOTRYOPHORUM, n. sp.

Frutex scandens, caulibus lignosis ferme unciam diametro; cortex suberosus albescens. Rami graciles glabri. Folia valde variabilia membranacea nonnunquam integra ovata lanceolata acuta petiolata, circiter 2 uncias longa, unciam lata, sæpius triloba, lobus medius ovato-lanceolatus 2–3 uncias longus, unciam latus, lateralia breviora falcata; haud raro præsertim in plantis juvenibus multilobata vel runcinata. Paniculæ nutantes multifloræ, ramosæ densæ pulcherrimæ. Flores eis *S. Seaforthiæ* subæquales, violaceæ, conniventes; pedicellis ¾ unciam longis. Calyx viridis parvus, dentibus parvis 5. Corolla ¼-uncialis, lobi 5, lanceolatæ acutæ, marginibus pubescentibus. Stamina 5, filamentis brevibus basi incrassatis, ad apicibus acuminatis; antheræ flavæ conicæ, conniventes, poris magnis duobus dehiscentes. Stylus gracilis curvulus; stigma capitatum parvum. Ovarium globosum. Baccæ parvæ globosæ coccineæ.

This beautiful plant was very abundant in the Sapate, and

also occurred on the east hills. In the woods it climbed over the shrubs, and formed almost impenetrable barriers. The stem at the base was often strong and thick, and covered with a corky bark. The leaves are very variable in shape, usually glabous; but in seedlings the margins are often edged with fine white hairs. The flowers form large pendent masses, and are of a fine violet colour, with a paler stripe in the centre of each petal; they never seem to open wide. The berries are about the size of those of *Solanum Dulcamara*, but globose and of a bright scarlet.

The plant is allied closely to *S. Seaforthiæ*, Andr.

SCROPHULARIACEÆ.

SCOPARIA DULCIS, *L. Sp. Pl.* ed. 1, p. 116; *DC. Prodr.* x. p. 431; *J. A. Schmidt in Mart. Fl. Bras.* viii. 3. p. 264.

Common on the main island, especially between the village and Fort San Antonio. Also very abundant on Rat Island. It is called here "Vassorinha," and reputed as a medicine for consumption.

Distribution. All the tropical world.

S. PURPUREA, n. sp.

Herba erecta, radice crassa lignosa, ramis erectis. Folia sub-verticillata, juniora angusta lanceolata integra, adulta cuneata runcinata basi angustata, dentibus acutis. Flores parvi 2–3 in axillis, pedicellis decurvis .... Calyx 4-fidus, sepala lanceolata acuta. Petala 4, ovata subobtusa vel lanceolata acuta, roseo-purpurea, pilis perpaucis. Stamina 4, filamenta glabra. Antheræ ellipticæ oblongæ, loculis basi disjunctis. Ovarium conicum; stylus cylindricus; stigma capitatum. Capsula globosa sepala vix superans. Semina minuta oblonga, atro-brunnea reticulata.

Rat Island; on the north side by the sea-shore, a few plants.

The colour of the flowers, pale rose, and almost complete absence of hairs from the base of the petals, so conspicuous in *S. dulcis*, distinguish this species from that; the habit is more erect and stiff, and the leaves larger and more toothed.

BIGNONIACEÆ.

BIGNONIA ROSEO-ALBA, n. sp.

Frutex altus, 15-pedalis vel ultra, caulis diametro ad 6-unciali, ramis strictis nec scandens, cortice brunneo verrucoso alabastris nigris. Folia trifoliolata, foliola ovata acuta vel ovata lanceolata superne glabra obscure viridia, subtus griseo-pubescentia, 5 uncias longa, 2½ uncias lata, petiolus ½-uncialis angulatus. Flores terminales pauci fugaces, pulchri. Calyx bilabiatus, ½-uncialis, lobus inferior bilobus, omnes oblongi obtusi, marginibus pubescentibus brunneis. Corolla tenuis sub-bilabiata 2 uncias longa, tubo dilato, lobi 5, lati rotundati, rosea et alba, signis flavis in labio inferiore, marginibus minutissime ciliatis. Stamina 5, filamentis gracilibus, basi hispidis. Antheræ versatiles arcuatæ angustæ brunneæ; appendice minima. Pollen albescens. Stylus gracilis albus. Stigma truncatum emarginatum. Ovarium teres purpurascens, discus parvus annularis viridis. Fructus immaturus gracilis teres.

This shrub, almost attaining the dimensions of a tree, is very abundant in the Sapate; also on Look-out Hill; but is very much sought for firewood, so that large plants of it are not very common. Its flowers are borne in the axils of the upper leaves, and are very thin and fugacious. The corolla is rose-colour and white, irregularly mingled with some yellow markings in the lower lip and throat. The buds are black, like those of an ash-tree. The fruit we could only procure immature, inasmuch as the plant flowered only towards the end of our stay in the island. It seemed to be slender, cylindrical and curved.

VERBENACEÆ.

LANTANA LILACINA, *Desf. Cat. Hort. Par.* ed. 3, p. 392; *Schau. in Mart. Flor. Bras.* ix. p. 261, t. xliv. fig. 1.—*L. fucata, Lindl. Bot. Reg.* t. 798.

In and about the gardens of the convicts. I never saw it wild.

L. AMŒNA, n. sp.

Frutex mediocris, ramis tetragonis hispidis præsertim versus apices gracilibus. Folia ovata acuta denticulata, in petiolo decurrentia undique cinereo-pubescentia aromatica, lamina 2 uncias longa, 1½ uncias lata quo latissima. Capitulæ parvæ globosæ, ¼ unciam longa. Bracteæ lanceolatæ acutæ pubescentes. Calyx brevis bilobus pubescens. Corolla alba fauce flava ¼-uncialis, tubo gracili curvo extus glanduloso pubescenti, lamina bilabiata, labio superiore trilobo, lobis rotundatis, labio inferiore rotundato latiore emarginato, fauce hispido. Stamina 4. Antheræ latæ oblongæ, filamenta brevia. Stylus crassus; stigmate laterali rotundato parvo. Ovarium rotundatum pubescens. Drupa minima pubescens, ferme exsicca, sicca valde aromatica.

In the thickets of the Sapate, tolerably plentiful. An infusion of the leaves is used as tea, as is the case with other species in Brazil. This species is allied closely to *Lantana cinerascens* in its dry nut, which, like the leaves, is deliciously aromatic.

LABIATÆ.

HYPTIS PECTINATA, *Poit. in Ann. Mus. Par.* vii. p. 474; *Benth. in DC. Prodr.* xii. p. 127.—Nepeta pectinata, *L. Sp. Pl.* ed. 2, p. 799.

Abundant in the Sapate, where it grows to a height of about 6 feet. Flowers small, lilac.

Distribution. All parts of warm America. Also occurs in Africa and India.

II. SUAVEOLENS, *Poit. in Ann. Mus. Par.* vii. p. 472, t. 29. fig. 2; *Benth. in DC. Prodr.* xii. p. 126.—Ballota suaveolens, *L. Sp. Pl.* ed. 2, p. 815.—Bysteropogon suaveolens, *L'Hérit. Sert. Angl.* p. 17.

In open places on Tobacco Point, and also on the cliffs on Portuguese Bay, where the whole air was scented with it. It attains a height of from 4–5 feet. Flowers larger than the preceding, blue.

Distribution. Common in Tropical America, and occurring also in the East Indies.

PLANTAGINEÆ.

PLANTAGO MAJOR, *L. Sp. Pl.* ed. 1, p. 112; *J. A. Schmidt in Mart. Fl. Bras.* vi. p. 169.

This occurred as a garden-weed in the precincts of the Governor's house; and was no doubt accidentally introduced. It has some reputation here as a medicine in chest complaints. The plant seems to be scattered about on the coast-line of Brazil, from Bahia to Rio de Janeiro.

APETALÆ.

NYCTAGINEÆ.

BOERHAAVIA DIFFUSA, *Sw. Obs.* p. 10.—B. paniculata, *Rich. in Act. Soc. Hist. Nat. Par.* i. p. 105; *J. A. Schmidt in Mart. Fl. Bras.* xiv. 2. p. 369.

A common weed in the garden, in waste places, and also on the roads near Sambaquichaba and in the Sapate. Flowers bright pink. Also on Rat Island.

Distribution. Common all over the tropics.

B. HIRSUTA, *Willd. Phyt.* i. p. 23; *J. A. Schmidt in Mart. Fl. Bras.* xiv. 2. p. 370.

A specimen gathered by Moseley during the 'Challenger' Expedition seems to belong rather to this species. It was probably obtained in the village.

PISONIA DARWINII, *Hemsl. Voy. 'Chall.' Exped., Atlantic Isles,* p. 20, pl. xlvii.

This was described and figured from a poor specimen obtained by Charles Darwin during the voyage of the 'Beagle;' but the material obtained appears to have been so bad that both the description and figure are very misleading; therefore I think it will be more satisfactory to redescribe it.

Frutex altus vel arbor parva, foliis nitide viridia ovata vel ovata lanceolata acuta vel acuminata, superne glabra (sicca nigra) subtus pubescentia, lamina 4–7 uncias longa, 2–4½ uncias lata, petiolo rigido unciali; folia juvenilia cum ramis omnino ferrugineo-pubescentia. Flores congesti in capitulis parvis in apicibus ramis minimis, omnino ferrugineo-pubescentes. Perianthium tubulosum, lobis 5 brevibus acutis. Stamina.... Filamenta gracilia brevia, perianthium non superantia. Antheræ rotundatæ obtusæ, albæ. Stylus teres perianthio subæqualis, stigmatibus capillaceis ramosis. Ovarium conicum. Fructus dependens, oblongus, viridis, ½-uncialis, perianthio persistente coronatus striatus; sicca nigra.

This shrub or small tree is very common all over the main island. It attains a considerable size, and is conspicuous from its glossy green leaves; but although so abundant, we only found a single tree in bloom. This, however, bore a plentiful supply of flowers and two fruits. The stems are covered with a grey bark; but the younger branches are thickly coated with a rusty-brown pubescence, which overlies not only the twigs but also the young leaves and inflorescence. The flowers are closely crowded into small heads at the ends of the branches and are very small. The perianth is red, with the rusty pubescence outside, glabrous within; the lobes short and acute. The stamens are represented in the plate above quoted as projected far beyond the perianth-tube; I never saw them like this. They are slender and short. The pistil is terminated by a tuft of filaments at the apex, as in most other species; these are omitted altogether in the plate, in which the pistil is mutilated. The fruit is oblong, slightly pubescent, and green when fresh, and marked with low ribs, and terminated by the remains of the perianth-tube: it hangs down when ripe.

AMARANTACEÆ.

AMARANTHUS CAUDATUS, *L. Sp. Pl.* ed. 1, p. 990, ed. 2, p. 1406.
A few scattered plants occurred near the village, in Peak Bay, and also in some of the gardens.

A. GRACILIS, *Desf. Tabl. Hort. Par.* ed. 1, p. 43.—Euxolus caudatus, *Moq. in DC. Prodr.* xiii. 2. p. 274.
Peak Bay, and also by the roadside in the central district.
Distribution. World-wide.

A. VIRIDIS, *L. Sp. Pl.* ed. 2, p. 1405.—Euxolus viridis, *Moq. in DC. Prodr.* xiii. 2. p. 274.
Along the paths through the Sapate. Common.
Distribution. World-wide.

PHILOXERUS VERMICULARIS, *R. Br. Prodr. Fl. Nov. Holl.* i. p. 410.—Iresine vermicularis, *Moq. in DC. Prodr.* xiii. 2. p. 340. —Var. AGGREGATA: Iresine aggregata, *Moq. l. c.*—Philoxerus aggregatus, *H. B. K. Nov. Gen. et Sp.* ii. p. 203.
This plant is exceedingly common along the shores of the main island, Rat Island, &c. It covers densely the fallen boulders on the hill-slopes just above the beach. The common form was the

variety *aggregata*; but in Chaloupe Bay, where it was very luxuriant, there was a form with conical heads nearly an inch in length, which is doubtless the variety *longespicata* of Scubert in Mart. Fl. Bras. v. p. 226.

Distribution. Tropical South America and Africa.

CHENOPODIACEÆ.

Chenopodium anthelminticum, *L. Sp. Pl.* ed. 1, p. 220 ; *Moq. in DC. Prodr.* xiii. 2. p. 73 ; *Fenzl in Mart. Fl. Bras.* v. p. 150.

Cultivated as a medicinal plant in many of the gardens, and escaped thence in several places.

Basella alba, *L. Gen. Pl.* n. 382 ; *Jacq. Eclog.* t. 161.

Occurred as a garden-weed in the Governor's garden and also in the village.

PHYTOLACCACEÆ.

Rivina lævis, *L. Mant.* p. 41 ; *Lam. Ill.* t. 81. fig. 2 ; *J. A. Schmidt in Mart. Fl. Bras.* xiv. 1. p. 335.

Common everywhere on the main island and also on Rat Island, where the plants formed shrubs. The flowers of this form were white, the berries at first orange-yellow, after red. On the slopes towards the sea at Tobacco Point was another form with reddish stems and pinkish flowers.

Distribution. Common all over Tropical America ; very plentiful at Pernambuco.

EUPHORBIACEÆ.

Euphorbia comosa, *Vell. Fl. Flum.* t. 15 ; *Boiss. in DC. Prodr.* xv. 2. p. 66 ; *Muell.-Arg. in Mart. Fl. Bras.* xi. 2. p. 693.

In open stony places, in Chaloupe Bay and near Tangle Rock. It is called "Alvelose," and used medicinally.

Distribution. Brazil only ? chiefly known from the south.

E. pilulifera, *L. Amœn. Acad.* iii. p. 114 ; *Boiss. in DC. Prodr.* xv. 2. p. 21 ; *Jacq. Ic.* i. t. 478 ; *Muell.-Arg. in Mart. Fl. Bras.* xi. 2. p. 184.—E. hirta, *L. Amœn. Acad.* iii. p. 114.—E. capitata, *Lam. Encycl.* ii. p. 422.

Abundant in the garden and village ; also at Sambaquichaba and near Tangle Rock. A common prostrate weed.

Distribution. World-wide.

EUPHORBIA THYMIFOLIA, *Burm. f. Fl. Ind.* p. 2 ; *Boiss. in DC. Prodr.* xv. 2. p. 47; *Ic. Thes. Zeyl.* t. 105 ; *Muell.-Arg. in Mart. Fl. Bras.* xi. 2. p. 684.—E. thymifolia, var. β, *L. Amœn. Acad.* iii. p. 115.

A common weed in the garden and village ; also in other half-cultivated spots.

Distribution. Cosmopolitan.

E. HYPERICIFOLIA, *L. Sp. Pl.* ed. 1, p. 454; *Hook. Exot. Fl.* i. t. 36; *Boiss. in DC. Prod.* xv. 2. p. 23.—E. cuspidata, *Bertol. Misc. Bot.* iii. p. 433, t. 22. f. 2.

On the sea-shore among stones, not common, below the Sapate near Cape Placellière. Rat Island, north side.

Distribution. West Indies, Mexico, Guatemala, Guiana.

MANIHOT UTILISSIMA, *Pohl, Pl. Bras.* i. p. 32, t. 24; *Muell.-Arg. in Mart. Fl. Bras.* xi. 2. p. 457, *et in DC. Prodr.* xv. 1. p. 1064.

The cultivation and preparation of the Cassava occupies much of the time of the convicts. A considerable quantity is grown for use in the island and for export. It does not appear to escape from cultivation, nor have I seen it here in flower.

RICINUS COMMUNIS, *L. Sp. Pl.* ed. 1, p. 1006; *Muell.-Arg. in Mart. Fl. Bras.* xi. 2. p. 420.

The common form here seems to be the var. *brasiliensis* of Muell.-Arg. It is very common in every part of the main island except the wooded districts ; and also occurs plentifully on Rat Island. It does not appear to be cultivated.

JATROPHA CURCAS, *L. Sp. Pl.* ed. 1, p. 1006 ; *Jacq. Hort. Vindob.* iii. p. 36, t. 63 ; *Muell.-Arg. in Mart. Fl. Bras.* xi. 2. p. 487.

There were a few shrubs of this species in the village, forming hedges with the other species. It is used in medicine.

J. POHLIANA, *Muell.-Arg. in Mém. Soc. Phys. Genèv.* xvii. 2, p. 499 ; *Boiss. in DC. Prodr.* xv. 2. p. 1091 ; *Muell.-Arg. in Mart. Fl. Bras.* xi. 2. p. 492.—Var. SUBGLABRA.

This is the plant collected by Moseley during the ' Challenger ' Expedition, and not *I. gossypifolia* as published, from which species it is very distinct. It is very abundant on Rat Island and the main isle, and is very conspicuous from its bluish-grey bare branches. It grows on the open parts of the whole island in the form of a branching bushy shrub, bare except at the apex of the branches. In the Sapate it attains a tree-like form and habit, with a stem 2 inches in diameter, and white bark like that of birch. When cut it exudes a slimy juice which stains linen permanently. The open flowers are not red in this form at least, but bright yellow, the petals tipped with red, and the buds red. The leaves are almost glabrous, but there is a fine pubescence along the edge. It is very common round Pernambuco, and forms hedges in the suburbs.

Distribution. Brazil.

JATROPHA URENS, *L. Sp. Pl.* ed. 1, p. 1007; *Muell.-Arg. in DC. Prodr.* xv. 2. p. 1100, *et in Mart. Fl. Bras.* xi. 2. p. 500; *Jacq. Hort. Vindob.* t. 21.

Common and widely scattered in Rat Island, Sella Giueta, and the main island. The form here corresponds with the var. *genuina* of Muell.-Arg. It has, however, less deeply cut and more rounded leaves with no teeth, and but slight sinuations along the edges. It is not the common form round Pernambuco. The whole plant, including even the apices of the petals, is covered with stinging hairs. It is a small shrubby plant about 2 feet high, with spreading branches and white flowers of medium size. I am informed by Senor Mendonça that good thread is obtained from this plant. It is called " Ortega branca " and " Cançançao " by the natives, who say that leaves or branches of this plant put in a mouse's hole will at once drive away the occupant.

Distribution. All warmer parts of America.

PHYLLANTHUS LATHYROIDES, *Muell.-Arg. in Linnæa*, xxxii. p. 42; *in DC. Prodr.* x. 2. p. 404; *in Mart. Fl. Bras.* xi. 2. p. 52.

Very common and variable in open spaces and waste ground, Rat Island, near the blowhole. Main island everywhere, except in the wooded districts. A form with variegated leaves occurred at Sueste. The common form here appears to be the var. *commutata*, Muell.-Arg., as the filaments of the stamen are but shortly free at the top.

This is the plant referred to *P. brasiliensis*, Muell.-Arg., in the Voyage of the ' Challenger,' p. 22.

CROTON ODORATUS, n. sp.

Frutex ramosus suaveolens, virgulta densa formans 4-6-pedalis. Rami lignosi, cortice griseo verruculoso tecti, apicibus

juniorum pubescentibus. Folia ovata acuta, dentata, palmi-
nervia, undique parce stellato-hispida præsertim in venis, lamina
2 uncias longa, 1½ uncias lata, petiolus patulus, 1½ uncias longa.
Stipulæ breves integræ lanceolatæ, basi glandulosæ pubescentes.
Racemi terminales singuli erecti, rhachide pubescente, 4-unciales.
Flores feminei pauce dissiti ad basin racemi 3–4, masculi plures
congesti, feminei remoti. Flores masculi, pedicelli breves dense
albo-pubescentes. Sepala 10, ovata obtusa pubescentia. Petala 5,
lanceolata, linearia obtusa tenuia alba, filamentis æqualia. Stamina 9, filamenta graciles ferme nudi. Antheræ rotundatæ.
Discus pubescens, glandulæ pulvinatæ rotundatæ. Flores feminei
majores, pedicellis ut in masculis. Sepala oblonga lanceolata
viridia pubescentes inæqualia non glandulosa, 5. Discus glandulis factus rufus. Petala nulla. Ovarium subtriangulatum
omnino alba pubescentia tectum. Styli 3, brachia in utro 4
rufescentia. Capsula parce pubescens. Semina oblonga, marginibus obtusis, atro-brunnea undique minute punctata ; caruncula
alba.

A bush of considerable size in the wooded districts, in more
open parts usually about 4 feet high, but sometimes 10 feet in
height, the stems 2 inches through ; forming dense thickets very
difficult to penetrate. Very common on the main island, especially in the west at Tangle Rock and in the Sapate ; a thicket
of it occurred above S. Antonio Bay, and it also occurred on
Rat Island.

The wood is compact, hard, and white. The leaves are very
pleasantly aromatic when bruised ; they are of a light green,
turning bright red on withering. The plant is allied to *Croton
populifolius* of the West Indies.

ACALYPHA NORONHÆ, n. sp.

Suffrutex pedalis, caulibus juvenilibus pubescentibus. Folia
ovata lanceolata acuminata crenato-dentata, parce hispida, lamina
ad 4 uncias longa, 2 uncias lata ad basin, petiolus 3-uncialis ;
stipulæ lanceolatæ acuminatæ pubescentes. Racemi plures in
axillis foliorum 2-unciales, floribus femineis paucis dissitis ad
basin cæteris masculis congestis. Bracteæ femineæ ferme integræ, cordatæ, marginibus ciliis glandulosis munitis ; stipulæ
lineares lanceolatæ pubescentes, bractea haud superantes. Perianthium. . . . . Ovarium sessile, obtuse trigonum dense albo-
pubescens ; styli valde ramosi tenues kermesini. Capsula parva,

hispida. Semen ellipticum brunneum punctatum. Flores masculi plurimi, globosi breviter pedicellati, in capitulis munitis congestis. Bracteæ breves, flores vix superantes, laciniatæ. Sepala
4 ovata obtusa. Stamina 5, filamentis brevibus. Antheræ
arcuatæ, loculi recurvi, discus parvus.

On the slopes of the Peak, among the boulders. A small
shrubby plant, tolerably plentiful at this spot, but not seen
elsewhere.

TRAGIA VOLUBILIS, *L. Sp. Pl.* ed. 1, p. 980; *Muell.-Arg. in
DC. Prodr.* xv. 2. p. 935.

We did not find this plant during our visit; but some fruits
and leaves were sent afterwards. It is termed " Ortega trepa-
deira " and " Tamiarana ; " and reported to be so poisonous that
any animal eating it among other herbage speedily dies.

Distribution. West Indies, Brazil, and Peru.

SAPIUM SCELERATUM, n. sp.   (Plate III.)

Arbor magna ad 30-pedalis, ramosa, valde laticifera, cortice
griseo. Folia iis *Pruni Laurocerasi* simulantia tenuiora, lanceolata, marginibus dentatis, glandulis parvis conicis ad basin
laminæ et rarius in marginibus ; lamina atro-viridis nitens ad
5 uncias longa et 2 uncias lata, petiolus uncialis cum glandulis
rufus. Stipulæ brevissimæ ovatæ. Racemi 1½-unciales, in apicibus ramorum foliis denudatorum, rhachide crassiusculo, floribus
femineis 1–2 ad basin, masculis pluribus remotiusculis. Flores
feminei : glandulæ 2 oblongæ, ad basin sæpe sepala minuta ovata 3,
ferme cœlantes. Pistillum conicum crassum. Styli rufescentes,
recurvi, validuli. Flores masculi plures 4 congesti, glandulis
duabus ut in femineo. Sepala 2, oblonga ovata obtusa viridia,
apicibus roseis. Stamina 2, filamentis basi subincrassatis, apice
attenuatis. Antheræ conicæ flavæ. Capsula parva globosa,
bivalvis, ⅜ unciam longa, crassiuscula, columella persistens. Semen
unicum ovatum griseum, ¼ unciam longum, basi rotundata, apice
acuto, uno latere complanato.

This plant, known as the " Burra," occurs on all the islands of
sufficient size—Rat Island, Sella Giueta, and all parts of the
main island. Although mentioned by Webster under the name
of the laurelled Bara, and alluded to by Darwin and Moseley,
specimens do not seem ever to have been brought to this country,
at least adequate for description ; indeed, no one seems to have

seen the flowers: flowering specimens were by no means easy to obtain, as the plant had hardly any flowers when we left the island; they only appear on trees which have shed their leaves at the approach of the hot season. The Burra is one of the largest trees on the island, attaining a height of about 30 feet and a very considerable thickness. It has wide spreading branches, which in old trees are but thinly covered with leaves. The bark is smooth and grey. Every portion of the plant, except the wood, exudes when wounded an abundant white latex of very acrid nature. This is so poisonous, that it is said to burn off the hair of horses and cattle where it touches the skin; and care is taken not to tie a horse up to a burra-bush. As the twigs are very brittle, persons pushing through a bush are liable to get the milk thrown in the eye, when it is stated to cause blindness. Mr. Lea met with this accident on one occasion on Sella Giueta, a drop of the milk entering one of his eyes, creating a bad inflammation which lasted for some hours. Human milk and urine are recommended as lotion in such an accident. Some of the convicts planted hedges of it round their gardens in order to deter thieves from breaking in at night.

The leaves of the plant resemble those of the Portugal Laurel, which resemblance is increased by the strict habit of the branches of the young plants. They are deep shining green, with a red petiole, and rather thin in texture, unlike those of *Sapium biglandulosum*. On the upper part of the petiole are two little conical red glands; and similar glands occur also not rarely upon the margins of the blade. The racemes are very short, not more than $1\frac{1}{2}$ inch long, with one or two sessile female flowers at the base, the upper portion being covered with the male flowers in little clusters of four. The female flowers are single, and just below them are a pair of oblong succulent pinkish glands. The perianth of three lobes is very minute, and often hardly visible on account of the glands; the pistil is green and conical, but constricted towards the apex; the styles are recurved, thick and red. Below each cluster of male flowers is a pair of glands like those of the female flower. There are usually four flowers in a cluster, opening one after another; each consists of a pair of small perianth-segments, alternating with which are a pair of stamens. The capsules are small, subglobose, and bluntly 3-angled; and contain a single grey seed a quarter of an inch long, broad and rounded at the base and more acute at the apex.

Although the seeds are so acrid and poisonous, yet I am informed by the Director of the island that the small birds eat them largely, and pass them unchanged. "So when it rains we meet with little burra-trees, which are cultivated. Thus it is well seen that such birds should be very well able to cover in a short time the whole island with burra-trees, as it was once, if it was not inhabited any more." This accounts for the diffusion of the plant into every corner of the island.

URTICACEÆ.

FICUS NORONHÆ, *Oliver in Hook. Ic. Plant.* xiii. t. 1222, p. 18; *Hemsl. Voy.* 'Chall.,' *Bot. Atlantic Isles,* p. 23.

This Fig-tree, peculiar to this group of islands, was partially described and figured in the above-mentioned place from material obtained by Moseley. The description is, however, incomplete from poverty of material; and therefore I think it well to redescribe it.

Arbor magna, radicibus acriis copiosis longissimis, ad 15-pedalibus, cortice griseo lævi, in plantis junioribus, in vetustis rimoso, valde lacticifero. Folia coriacea atro-viridia nitida, elliptica obovata obtusa, 4–9 uncias longa, 2½–4 uncias lata, petiolus incrassatus, ½–1 unciam longus. Receptacula globosa unciam in diametro, viridia, maculis purpureis vel omnino purpurea, bracteis inferioribus 3, ovatis obtusis. Flores masculi et feminei undique commixti in receptaculo. Bracteæ lanceolatæ dentatæ, floribus æqualibus. Flores masculi quam feminei pauciores, stipitati vel sessiles, bracteolis duabus ad basin. Perianthium trilobum, lobi ovati obtusi. Stamen unicum, filamento crasso. Anthera terminali plana, oblonga, loculis approximatis. Flos femineus stipitatus vel sessilis, bracteolis 2 lanceolatis integris iis masculi similes. Perianthium trilobum, lobis ovatis obtusis. Stylus gracilis breviter e perianthii extrorsus bifidus, ramis brevibus recurvis. Ovarium ellipticum. Fructus ovalis. Semen ovale, 1 mm. longum, album.

Rat Island, Sella Giueta, main island.

This tree has been so extensively cut down, that but few of any size are now left in the islands. It attains its greatest dimensions in the Sapate; but does not grow in the thickest part of the woods, but in the more open spots. The finest trees now in existence are in the garden of the Director's house, where four very large specimens grow on the banks of the stream. On the cliffs, as at the Peak, and on very exposed spots, as at Rat Island, it is of much lower growth, and forms a shrub creeping over the rocks by means of its aerial roots, or springing from clefts. Small bushes of it grow in the highest and most inaccessible parts of the Peak, and on some of the smallest islands, as the Dois Irmaos. The roots are very long, slender and tough, of a light brown colour; and are used to make whips for chastising the convicts. The bark and leaves, when broken, emit much milk, are very sticky, and apparently contain a considerable amount of caoutchouc; this milk is used as bird-lime. The leaves are glossy dark green, and much infested by galls. The figs are rounded, and either green with purple spots, or entirely purple; they are about an inch through when ripe, and are sweet and of somewhat pleasant flavour; at their base are three short oval bracts. The bracts at the mouth of the receptacle are numerous, nearly all being inverted, the two everted ones forming a short conical umbo. The flowers of both sexes are irregularly mixed, the females being most numerous. Among them are many laciniate bracts; and each flower is subtended by two small bracteoles. Many of the flowers are supported on a stalk, others are quite sessile. The perianth seems to be similar in both the male and female flowers, consisting of three short overlapping oval blunt lobes. The style is short and bifid.

The plant is called " Gamaleira " here, as are other species of the genus in Brazil.

ARTOCARPUS INCISA, *Forst.*

Several Breadfruit-trees occur in the gardens and in various spots in the village. On one tree a male spike was found at the base of which were female flowers.

FLEURYA ÆSTUANS, *Gaudich. in Freyc. Voy. Bot.* p. 497.— Urtica æstuans, *L. Sp. Pl.* ed. 2, p. 1397; *Jacq. Hort. Schœnbr.* iii. t. 388.

This plant occurred sparingly on Tobacco Point, growing among *Philoxerus vermicularis* and *Canavalia*, and also by the lake. Specimens obtained in the latter locality were slightly stinging, and had very widely spreading panicles. It appears to be very variable; and specimens obtained at Pernambuco, where it is

common, differed somewhat in appearance from both the forms
met with here.

Distribution. All over tropical S. America.

MONOCOTYLEDONS.

There were no petaloid Monocotyledons in the island except a
few introduced by man ; and these had hardly established themselves. An *Hymenocallis*, an *Aloë*, *Zanonia discolor*, and *Furcræa
gigantea* occurred in the gardens, the latter cultivated for its
fibre, as elsewhere in Brazil. Bananas of several varieties were
largely cultivated, and sold for from six to nine for a vintem (about
a halfpenny). The largest and richest plantation was on the
slopes of the Peak.

PALMÆ.

COCOS NUCIFERA, *L. Sp. Pl.* ed. 1, p. 1188.

Cocoa-nuts are largely cultivated in the sandy bays at Sueste
and Sambaquichaba, and in one or two other spots. All the trees
on the island are the property of the Director, none of any large
size, and appear to be comparatively recent introductions. They
fruit very well, and are usually loaded with nuts. At Sueste was
a specimen with a branched stem ; the main stem had apparently
fallen forward from shifting of the sand, and had then thrown
up a second stem.

COPERNICIA CERIFERA, *Mart.*

Carnauba Palm. There were a few young plants of this near
the Peak, and a larger one in the village.

OREODOXA REGIA, *Kunth.*

The are two trees of this palm by the door of the church.

CYPERACEÆ.

CYPERUS CIRCINATUS, n. sp. (Plate II. figs. 1, 2.)

Pusillus, 2–5-uncialis, culmis pluribus flaccidis, vaginis paucis
papyraceis ad basin. Folia pauca angustissima linearia circiter
uncialia, culmis haud æquantia. Culmi triquetri 2–5-unciales,
umbellis simplicibus breviter 1-radiatus aut subcapitatus, spiculis
dissitis ½-uncialibus, apicibus circinatis; bracteæ 3 lineares acuminatæ, marginibus scabridis longissimis, circiter 3 uncias longæ,

et ferme $\frac{1}{16}$ unciam latæ. Spiculæ ad 20-floræ angustæ. Rhachis sinuata tenuis. Squamæ lanceolatæ subacutæ, flavæ; carina viridi. Stamen unicum. Stylus bifidus ruber, squamam vix superans. Caryopsis oblonga, apice breviter mucronato, testacea puncticulata obscure biconvexa.

This little *Cyperus* was only met with in clefts of rock on the Peak, and on the slopes of Morro branco. It is 2-styled, the nut being rather long and almost terete, showing very slight traces of biconvexity. The spikelets, which are slender and have the flowers rather distant, are curiously curled at the apex in most cases.

CYPERUS COMPRESSUS, *Presl, Rel. Hænk*. iii. p. 177; *Rottb. Gram.* p. 28, t. 9. fig. 3; *Boeck. Cyp. Herb. Berol.* p. 121.

Common on the main island, on the sea-shore in Peak Bay, and also among the stones in the roads near San Antonio and in the village, as far west as Leao Bay, where a very dwarf form with short erect spikelets was seen. The plants were all small, as if the species had only begun to establish itself in the island.

Distribution. All tropical countries.

C. VIALIS, n. sp.

Annuus; basi vaginis rufescentibus tecti. Folia linearia acuminata acuta, 5 uncias longa, 1 lineam lata. Culmi validuli triquetri 7-unciales. Umbella triuncialis 4–6 radiorum erectorum, spiculis pluribus in apicibus congestis, et nonnullis sessilibus in medio umbellæ. Bractæ 4, umbella subæquales, lineares acuminatæ. Spiculæ $1\frac{1}{2}$–$1\frac{3}{4}$ uncias longæ, graciles, 40–50-floræ, colore *C. compressi*. Squamæ lanceolatæ mucronatæ, marginibus late scariosis, carina viridi, lateribus sæpe rufescentibus. Stamina 3, filamentis gracilibus. Antheræ lineares apiculatæ. Stigmata 3, brevia rufa. Caryopsis oblonga angusta, trigona, angulis obtusis, testacea. Rhachis gracilis exalata.

Only two specimens of this plant were obtained, both at considerable distances apart along the roadsides, in the central district.

From its habitat I felt convinced it was an introduced plant, but have never been able to match it with any species. It is allied to *C. rotundus*, but the exceedingly long spikelets and form of the nut make it quite different. It has, too, the green and white colouring of *C. compressus*, and not the red glumes of *C. rotundus*. Mr. C. B. Clarke, to whom I showed the plants, pointed out the relationship to *C. rotundus* forma *viridis*, from which, however, he considers it distinct.

CYPERUS BRUNNEUS, *Sw. Fl. Ind. Occ.* p. 116 ; *Griseb. Fl. Brit. W. Ind.* p. 565.—C. purpurascens, *Vahl.*—C. atlanticus, *Hemsl. Bot. Voy.* 'Chall.,' *Atlantic Isles,* p. 130, t. xxiii.—C. olidus, *Rich.*

The commonest species of this genus upon the island is identical with the *C. atlanticus,* Hemsley, obtained by Sir Joseph Hooker upon the island of South Trinidad ; and I am quite unable to separate it from *C. brunneus,* Sw., of which there are type specimens in the British Museum. Swartz's specimens are small and stunted, but the plant grows much bigger and is very common throughout the West Indies, being apparently a sea-shore plant. I have seen it from Jamaica, Cuba, Tortola, St. Croix, Bahamas, Martinique, St. Bartholomew, Barbados, and it also occurs in Yucatan and Florida. Grisebach adds as synonym *C. insignis,* Kunth, based on Sieber's Trinidad plant, no. 7, which differs in the longer, more cylindrical nut, and narrower, more acute, almost unribbed glumes, and also in habit, and is the *C. planifolius,* Richard, as is shown by a type from St. Croix in the British Museum. *C. brunneus,* Sw., occurs on Fernando Noronha, on Rat Island, in the open country with *C. ligularis,* and on the main island at Leao Bay, Tangle Rock, Chaloupe Bay. In the sand of Portuguese Bay and on rocks at the end of the Sapate on Cape Placellière was a more glaucous form with pallid spikelets. In the open country it forms large tussocks very similar to those of *C. ligularis.* It varies greatly in breadth of foliage.

It is interesting to find this plant occurring in the West Indies and Oceanic islands, but not on the South-American continent. Perhaps it may be one of those plants which have been distributed by the drifting of their seeds in the sea.

C. NORONHÆ, n. sp.

Planta rupicola flaccida, rhizomate brevissimo, basi dilatato Folia plura, angusta linearia acuminata debilia, circiter 18 uncias longa, 1½ lineas lata ; carina et margine scabrida. Culmi subæquilongi triquetri. Umbella radiorum 7–10 valde inæqualium patula. Bracteæ involucri 5, longæ lineares acuminatæ. Radius longissimus uncialis, basi vagina castanea. Spiculæ ad 30 in apicem congestæ, circiter ¼ unciam longæ, ferrugineæ. Rhachilla crassiuscula, alis scariosis magnis. Spiculæ sessiles in discis elevatis rotundatis. Glumæ duæ ad basin spiculæ ; gluma infima lanceolata subacuta, superior latior obtusa. Glumæ alteræ lanceolatæ acutæ, ferrugineæ, carina viridis, costis elevatis 10. Stamina tria. Stylus longuisculus trifidus rufus.

Caryopsis obovata, basi angustata, obtuse trigona brevissime apiculata rufa punctata.

Main island, high up on the Peak and in clefts in the rock of Chaloupe Bay; no. 7.

CYPERUS DISTANS, *L. f. Suppl.* p. 103; *Jacq. Ic. Pl. Rar.* ii. p. 8, t. 299; *C. B. Clarke in Journ. Linn. Soc. Bot.* xxi. p. 144.

Plentiful by the edges of the roads near the Peak in the cultivated district.

Distribution. The whole of the tropical world.

C. FERAX, *Rich. in Act. Soc. Hist. Nat. Par.* i. p. 106; *C. B. Clarke in Journ. Linn. Soc. Bot.* xx. p. 295, xxi. p. 191.—Diclidium ferox, *Schrad.*; *Nees in Mart. Fl. Bras.* ii. 1. p. 54.

Common in the central district with the last, but much more abundant; Sambaquichaba, Leao, Tangle Bay, &c.

Distribution. Common in South America.

C. LIGULARIS, *L. Am. Acad.* v. p. 31; *Sp. Pl.* p. 70; *Boeck. Cyp. Herb. Berol.*; *C. B. Clarke in Journ. Linn. Soc. Bot.* xx. p. 196.

Tangle Bay, and the field below the curral near Cotton-tree Bay and Leao. Not so common as the last. Abundant on Rat Island; forming large tufts.

Distribution. All warm parts of America from Florida to Rio de Janeiro, also west Tropical Africa and Madagascar.

FIMBRISTYLIS DIPHYLLA, *Vahl.*

Scattered about on the main island, on the hill above Chaloupe Bay, at East Point near the curral, and at Leao, mostly along the paths, in dry spots. At Morro branco it was common and a good deal larger, growing among grass.

Distribution. All the tropical world.

SCIRPUS MICRANTHUS, *Vahl.*—Hemicarpha isolepis.

In the fields below the Peak and round San Antonio Fort, in waste ground.

Distribution. All tropical countries. It is a common weed in the garden paths in Pernambuco, and has doubtless been introduced accidentally into Fernando Noronha.

RHYNCHOSPORA MICRANTHA, *Vahl, Enum.*; *Boeck. Cyp. Herb. Berol.* p. 768.—Dichromena micrantha, *Kunth, Enum.* p. 278.

On the hill between Chaloupe Bay and S. Antonio Bay, on the paths through the maize-fields in the Sapate, and near the Peak.

Distribution. West Indies, Guatemala, Brazil, West Africa, and Teneriffe.

GRAMINEÆ.

PASPALUM ANEMOTUM, n. sp.

Herba dense cæspitosa. Folia copiosa, flaccida elongata, $2\frac{1}{2}$–3-pedalia vix $\frac{1}{4}$ unciam lata linearia acuminata striata, scabra; ore et marginibus vaginæ longe albo-ciliatis. Ligula brevis membranacea brunnea laciniata, laciniis rotundatis. Culmi bipedales erecti. Panicula nutans 6-uncialis, racemis circiter 20. Rhachis gracilis vix complanata. Flores per paria parvi pallidi longi. Pedicelli breves scabridi. Glumæ exteriores ovatæ obtusæ subtenues, internæ induratæ lanceolatæ. Palea indurata anceolata. Stamina 3. Antheræ castaneæ. Stigmata breviuscula atro-purpurea.

Abundant on the open ground behind Fort San Antonio, in the low ground near Tangle Rock, and at Morro branco. This is a large plant, forming thick tussocks in low-lying country; the leaves are numerous and long and narrow, the inflorescences few and rather compact, the racemes long and slender, the rhachis hardly flattened. The flowers numerous and white. It belongs to the same section as *P. virgatum*, but even in habit is distinct, the leaves and inflorescence being much narrower.

P. PHONOLITICUM, n. sp. (Plate IV.)

Herba rigida erecta, vix cæspitosa. Folia pauca erecta late linearia, culmo multo breviora, acuminata striata scabrida, 6 uncias longa, $\frac{1}{2}$ unciam lata. Ore et marginibus vaginæ albo lanatis. Ligula laciniata, laciniis rotundatis membranaceis. Culmus sesquipedalis ad inflorescentiam vaginis tectus. Panicula erecto-nutans 6-uncialis, racemis 10. Rhachis sinuata scabra vix complanata. Flores per paria in pedicellis brevibus pubescentibus quam in præcedente paullo majores. Gluma externa late ovata obtusa carinata cymbiformis . . . . plana elliptica obtusa; gluma interna indurata ovata obtusa cymbiformis minute striata. Palea elliptica obtusa striata. Antheræ flavæ. Styli atropurpurei.

On the altered phonolite of Morro branco, growing in clefts of the rock and on the slopes.

This species is allied closely to the preceding, but is distinguished at first sight by its habit; it does not form the large long-leaved tussocks of that species, but grows in small tufts with a few erect, stiff leaves, much shorter and broader than those of the other. The whole plant, too, is smaller and more condensed.

The flowers, however, are larger, and are ovate in outline, instead of being almost lanceolate elliptic, while the glumes are deeper and more blunt, and the fertile glume and palea harder in texture.

PANICUM SANGUINALE, *L.*, var. CILIARE, *Doell, Rhein. Fl.* p. 126.—P. ciliare, *Retz. Obs.* iv. p. 16; *Kunth, Enum.* i. p. 82.

Paths of the garden at the Director's House. This was a quite typical prostrate form, similar to the Indian. Another variety, with numerous short lanceolate dark green leaves, an inch long by a quarter broad, and with geniculate stems, grew on Tobacco Point, and a more tufted dwarf form on the slopes of Morro branco.

The variety appears to be rare in Brazil, though very common in the Old World.

P. SANGUINALE, var. DISTANS, *Doell in Mart. Fl. Bras.* vol. ii. p. 134.—Digitaria horizontalis, *Willd. Enum.* i. p. 92; *Roem. & Schult. Syst. Veg.* ii. p. 474.

This is the commonest form in South America, and was very plentiful and varied here. A quite typical form occurred on Cape Placellière and on the north side of Rat Island, and again along the paths in the village. A very large and hairy form grew on the cliffs at Chaloupe Bay. It was about three feet high, with a decumbent base, and the leaf-sheaths were thickly covered with long white hairs a quarter of an inch in length. The flower-spikes also had, in the lower part, a number of similar white hairs on the rhachis, and the spikelets were pubescent. This variety was obtained also by Glocker at Bahia (no. 216 of his collection).

P. BRIZOIDES, *Lam. Ill.* i. p. 470, no. 882; *Doell in Mart. Fl. Bras.* ii. p. 184.—P. appressum, *Lam. Ill.* p. 176, no. 929.

In a damp spot in the centre of the isle and along the streams at Leao and in Sponge Bay, and also very dense and plentiful round the Lake. This plant is known here as "Gramma."

Distribution. Cosmopolitan.

P. PLANTAGINEUM, *Link, Hort. Berol.* i. p. 206; *Kunth, Enum.* i. p. 92; *Trin. in Mém. Acad. Pétersb.* 1835, p. 242; *Doell in Mart. Fl. Bras.* ii. p. 186.—P. Leandri, *Trin. Sp. Gram.* xxviii. t. 335.

Along the paths in the central district and in the maize-fields. Only a few plants, evidently introduced in crops.

Distribution. Texas, Mexico, and Brazil, from Pernambuco to Rio de Janeiro, also Bolivia and Australia.

PANICUM NUMIDIANUM, *Lam. Ill.* p. 902 ; *Doell in Mart. Fl. Bras.* ii. p. 188.

In the swamp by the stream at Leao, and two or three large patches on the path in the Sapate. It is called "Capim de Planta," and is, without doubt, introduced more or less intentionally, as it is the best fodder-grass in the north of Brazil and constantly cultivated.

P. FUSCUM, *Sw. Prodr.* p. 23 ; *Fl. Ind. Occ.* i. p. 156.—P. fasciculatum, *Nees, Agr. Bras.* p. 151.

A few scattered plants in Chaloupe Bay, in rock-clefts. In the sand of the shore of Peak Bay, and more plentiful and larger along the path into the Sapate with the preceding.

Distribution. Southern North America, West Indies, and South America.

P. TRICHODES, *Sw. Prodr.* p. 24 ; *Fl. Ind. Occ.* i. p. 176 (ex parte).—P. capillaceum, *Lam. Ill.* i. p. 173 ; *Doell in Mart. Fl. Bras.* ii. p. 249.

Plentiful on the Peak and at Tangle Rock, and is almost the only herbaceous plant growing under the bushes of the Sapate. It grows, too, on Sella Giueta. This is the plant distributed under the name of *P. parvifolium*, Lam., from the 'Challenger' Expedition collections. The species is a very common woodland plant in Brazil and other parts of South America and the West Indies ; but there seems to have been some confusion as to its name. The earliest name seems to be *P. trichodes*, Sw., which was based on a plant collected by Sir Hans Sloane in Jamaica, which is preserved in the Natural History Museum. In his later work, Swartz added to the species Linnæus's *P. brevifolium*, based upon a Ceylonese plant and quite distinct.

P. COLONUM, *L. Sp. Pl.* ed. ii. p. 84 ; *Jacq. Eclog.* t. 32 ; *Doell in Mart. Fl. Bras.* ii. p. 140.

Paths on Rat Island ; a purple-flowered form is common on the sandy shores of Peak Bay, and the common pale-coloured form grows all over the paths and waste ground in various parts of the main island near the village.

Distribution. A weed of world-wide distribution.

SETARIA SCANDENS, *Schrad.*; *Roem. & Schult. Syst.* ii. p. 279.
—Panicum scandens, *Trin. Diss.* ii. p. 166; *Doell in Mart. Fl. Bras.* ii. p. 170.

Common in thickets on the Peak and in Chaloupe Bay, scrambling over the lower plants, also on Rat Island.

Known here as " Carapicho." There was a very tall and stout erect or suberect form, 2½ feet high, with thick stems, growing in clefts of the rocks at Cape Placellière.

Distribution. South America, especially plentiful in Brazil.

The heads adhere tightly to clothes, &c., by means of the recurved processes on the setæ, and in this way the plant gets carried about from place to place.

S. CAUDATA, *Roem. et Schult. Syst.* ii. p. 495.—Panicum caudatum, *Lam. Ill.* no. 893; *Doell in Mart. Fl. Bras.* ii. p. 161.

This is chiefly a woodland plant, forming in the Sapate in open spots large tussocks very like those of *Brachypodium sylvaticum*, the foliage being very plentiful and dark green. It was plentiful in the Sapate, at Tangle Rock, and on the Peak. On Sella Giueta and also on the Look-out Hill there was a smaller half-prostrate form, with geniculate culms.

The plant is called " Capinche."

Distribution. Mexico to Brazil.

CENCHRUS VIRIDIS, *Spreng. Syst.* i. p. 301; *Doell in Mart. Fl. Bras.* ii. p. 309.

Among *Æschynomene hispida*, on the slopes towards the sea on the northern side of Rat Isle.

Distribution. S. America.

C. ECHINATUS, *L. Sp. Pl.* ed. 2, p. 1150; *Doell in Mart. Fl. Bras.* ii. p. 310, t. xliii.—C. pungens, *H. B. K. Nov. Gen. et Sp.* i. tab. xliv.

On the shores of the sandy bays at Peak Bay, Portuguese Bay, Leao, Chaloupe Bay, S. Antonio Bay, and in the Sapate; common.

Distribution. Whole tropical world. A very troublesome weed, on account of its prickly burr-like spikelets.

ZEA MAYS, *L.*

Is extensively cultivated here, and the grain exported to Pernambuco. The husking of the maize and preparing it for export are done by the convicts. One plant we saw had no less than nine full-sized cobs upon it.

ORYZA SATIVA, *L.*

Rice is cultivated to a small extent at Sambaquichaba.

ANTHEPHORA ELEGANS, *Schreb. Beschreib. Graes.* ii. p. 105; *Doell in Mart. Fl. Bras.* ii. p. 313.—Tripsacum hermaphroditum, *L. Sp. Pl.* ed. 2, p. 1378.—Cenchrus lævigatus, *Trin. Fund. Agrost.* p. 172.

Tolerably plentiful on the sandy ground behind Fort San Antonio, also near the Sapate.

Distribution. Tropical America.

A common weed in the grass-plots in Pernambuco.

SACCHARUM OFFICINARUM, *L. Sp. Pl.* ed. 1, p. 54.

The sugar-cane is largely cultivated in the main island.

ANDROPOGON SCHŒNANTHUS, *L.*?

A tuft of a very sweet-scented grass was found in a corner of a maize-field, where I believe it had been planted. There was no flower on it, but it apparently belonged to this species.

SORGHUM VULGARE, var. SACCHARATUM?

A few plants occurred in a maize and sugar-cane field near Sambaquichaba.

S. HALEPENSE, *Pers. Syn.* p. 101; *Doell in Mart. Fl. Bras.* ii. 2. p. 272.—Holcus halepensis, *L. Sp. Pl.* ed. 2, p. 1047.

A weed in the garden of the Director's house, and forming dense thickets on the slopes of Peak and Water Bay below the village. Also found on Rat Island.

ELEUSINE INDICA, *Gaertn. Fruct.* i. p. 8.—Cynosurus indicus, *L. Sp. Pl.* ed. 2, p. 106, no. 8; *Doell in Mart. Fl. Bras.* ii. 2. p. 86, t. xxiv.

A weed in the garden, and also exceedingly abundant among the stones of the paths throughout the main island. It seems to be the only plant that forms anything like turf here. It also grows on Rat Island.

Distribution. All warm countries.

E. CRUCIATA, *Lam. Ill.* t. 48. fig. 2.—Dactyloctenium ægyptiacum, *Willd. Enum.* ii. p. 1029; *Doell in Mart. Fl. Bras.* ii. p. 38, t. xxv.

Three forms of this common plant occur here:—The typical suberect form, with the leaves almost glabrous; abundant in waste ground in the village, &c., also on Rat Island. A tall, erect, but weak form, with flaccid, very hispid leaves; a single plant on Platform Island. And a flat prostrate plant, with short broad leaves and short blunt spikes, which is very plentiful on the roads near Sambaquichaba and Fort San Antonio. Doell, in the 'Flora Brasiliensis,' describes the leaves of this species as "glabra vel glabriuscula"; but most of the Brazilian specimens I have seen have distinctly hairy leaves with numerous white cilia.

Distribution. The whole of the warm parts of the world.

CHLORIS BARBATA, *Sw. Fl. Ind. Occ.* i. p. 200; *Doell in Mart. Fl. Bras.* ii. pt. 3, p. 67; *Steud.* p. 204.

Common grass, growing between the stones of the walls and paths, also plentiful in the Sapate and in Chaloupe Bay, where it attains the height of five feet.

Distribution. World-wide.

C. VIRGATA, *Sw. Fl. Ind. Occ.* i. p. 203; *Roem. & Schult. Syst.* ii. 6. p. 8; *Doell in Mart. Fl. Bras.* ii. p. 65, t. xviii.

Along the paths in the Sapate. Also as tall as the preceding and growing with it in Chaloupe Bay. In Sponge Bay there was a smaller prostrate form.

Distribution. West Indies and Brazil.

GYMNOPOGON RUPESTRE, n. sp.

Herba rigidula ad 1½-pedalis. Culmi plures graciles ferme ad inflorescentiam foliati. Folia dissita linearia erecto-patula, 4–5 uncias longa, ⅛ unciam lata, acuminata pubescentia. Vaginis 1½-uncialibus, ligula et marginibus vaginæ albo ciliatis. Rhachis tenuis breviuscula, spicæ dissitæ tenues ad basin albo pubescentes, ferme 3-unciales, pallidæ. Glumæ inferiores angustæ lineares acutæ, scabridæ. Gluma fertilis ad basin pubescens, pilis albidis, anguste lanceolata. Arista longa scabra. Gluma terminalis angustissime linearis, caryopsis linearis brunnea. Rhachilla producta scabrida, flore sterili gluma vacuo, basi pilosa, sistente.

This grass was tolerably abundant in Portuguese Bay, growing on the cliffs in tufts with *Hyptis suaveolens.* It also occurred along the woodcutters' track in the Sapate, and on the summit of Tangle Rock.

ERAGROSTIS CILIARIS, *Link, Hort. Berol.* p. 192.—Poa ciliaris, *L. Syst.* ed. 10, p. 82, n. 590.

Plentiful along the paths and on the rocks near the village, and at Leao, at the Sapate, and on the Peak. Also on the paths in Rat Island.

Distribution. All warm countries, but especially abundant in Brazil.

ARUNDO DONAX, *L.* ?

An introduced grass at Leao, without flowers.

GUADUA LATIFOLIA, *Kunth, Syn.* i. p. 254 ?

What appears to be this common Brazilian Bamboo has been introduced here, in a few places, but it was not in flower.

CRYPTOGAMIA.

PELLÆA GERANIÆFOLIA, *Fée, Gen. Fil.* p. 130; *Hook. & Bak. Syn. Fil.* p. 146.—Pellæa concolor, *Bak. in Mart. Fl. Bras.* i. p. 396, t. xliii. fig. 3.

Under rocks in damp spots on the Peak and on Tangle Rock. This was the only Fern seen, and it is very local. The dryness of the climate is evidently unsuited for ferns.

CHARACEÆ.

NITELLA CERNUA, *A. Br., in Monatsber. Berl. Akad.* 1858, p. 354.

The lake was almost entirely filled with this beautiful *Nitella,* which formed a solid mass. I am indebted to Mr. H. Groves for identifying it.

Distribution. Caraccas.

MUSCI.

By A. GEPP, M.A., F.L.S.

CALYMPERES RICHARDI, *C. Muell.*

Hab. On bark in the Sapate; fruiting specimen on rocks above Point Noir. Also obtained by Darwin ('Beagle' Expedition).

TORTULA, sp.
Hab. Cleft in the Peak.

TORTULA, sp.
Hab. On the ground, Morro branco.

HYPNUM, sp.
Hab. On a stone in the Sapate.

HYPNUM, sp.
Hab. On a stone in the Sapate.

HEPATICÆ.

RICCIA RIDLEYI, *Gepp,* n. sp.

R. fronde solida crassa dichotoma, laciniis obovato-oblongis canaliculatis, margine squamoso subtus membrana purpurea obsito, squamis imbricatis rotundatis ultra marginem paulum exstantibus purpureis margine pallido.

Hab. loco humido umbroso Tangle Rock, Insula Fernando Noronha (*H. N. Ridley*, Aug. 1887).

Dimensions of frond, 5 × 2; spore 0·1 × 0·6 mm.

This species falls into the scale-bearing section of Bischoff's subgenus *Lichenoides*. It most resembles *R. limbata*, Bischoff, but differs in the less acute, non-ascending margin of the thallus, the colourless border of the scales, and the direction, which is at right angles to the margin of the thallus. The scales spring from the purple membrane, which covers the underside of the thallus for about one half of the distance from the margin to the median line, and exhibits obscure transverse folds. The scales are expanded laterally so as to overlap one another in an imbricate manner. The laciniæ are emarginate, devoid of papillæ, and still united to the withered main thallus; they bear the fruit beneath the upper surface in the median groove. Several archegonia, but only one ripe fruit has been observed. Several minute closely involute thalli occur beneath the older plants. The latter are aggregated together, and are very firmly attached to the soil by innumerable short hairs.

ALGÆ.

By GEORGE M. MURRAY, F.L.S.

The Algæ collected by the 'Challenger' Expedition at Fernando Noronha and recorded by the late Professor Dickie (Linn. Soc. Journ. Bot. vol. xiv. p. 363) have been included for the purpose of completing the enumeration of Algæ from this island.

FLORIDEÆ.

CENTROCERAS CLAVULATUM, *J. Ag.*
Ridley, Lea, and Ramage!
Geogr. Distr. Throughout tropical and subtropical seas.

GIGARTINA TEEDII, *Lam.*
Ridley, Lea, and Ramage!
Geogr. Distr. Atlantic and Mediterranean.

CHRYSYMENIA ENTEROMORPHA, *Harv.* ?
'Challenger.'
Geogr. Distr. West Indies.

PEYSSONNELIA DUBYI, *Crouan.*
'Challenger.' *Ridley, Lea, and Ramage*!
Geogr. Distr. Atlantic.

GRACILARIA MULTIPARTITA, *Clem.*
'Challenger.'
Geogr. Distr. Tropical and subtropical Atlantic, Mediterranean, Indian Ocean, N. Zealand.

G. ARMATA, *Ag.* ?
'Challenger.'
Geogr. Distr. Mediterranean.

GALAXAURA CYLINDRICA, *Lamx.*
'Challenger.' *Ridley, Lea, and Ramage*!
Geogr. Distr. West Indies and Red Sea.

G. RUGOSA, *Lamx.*
'Challenger.' *Ridley, Lea, and Ramage*!
Geogr. Distr. Tropical Atlantic and Indian Ocean.

G. LAPIDESCENS, *Sol.*
Ridley, Lea, and Ramage!
Geogr. Distr. Tropical Atlantic, Pacific and Indian Oceans.

G. OBLONGATA, *Lamx.*
'Challenger.'
Geogr. Distr. Tropical Atlantic.

ZANARDINIA MARGINATA, *J. Ag.*
'Challenger'! *Ridley, Lea, and Ramage*!
Geogr. Distr. Tropical Atlantic, Indian Ocean, Australia.

LAURENCIA PAPILLOSA, *Forsk.*
Ridley, Lea, and Ramage!
Geogr. Distr. Throughout tropical seas.

L. SCOPARIA, *J. Ag.*
Ridley, Lea, and Ramage!
Geogr. Distr. Tropical Atlantic (America).

ACANTHOPHORA THIERRII, *Lamx.*
'Challenger.' *Ridley, Lea, and Ramage*!
Geogr. Distr. Tropical Atlantic.

A. MULTIFIDA, *Lamx.*
Ridley, Lea, and Ramage!
Geogr. Distr. Tropical Atlantic.

AMANSIA DUPERREYI, *Ag.*
'Challenger.'
Geogr. Distr. Martinique.

LITHOTHAMNION MAMILLARE, *Harv.*
'Challenger'! *Ridley, Lea, and Ramage*!
Geogr. Distr. Atlantic (Bahia to Tierra del Fuego), Africa (Cape Verde).

L. POLYMORPHUM, *Aresch.*
Ridley, Lea, and Ramage!
Geogr. Distr. Throughout all seas.

JANIA CUBENSIS, *Mont.*
'Challenger'!
Geogr. Distr. West Indies.

J. RUBENS, *Lamx.*
Ridley, Lea, and Ramage!
Geogr. Distr. Throughout all seas.

CORALLINA CERATOIDES, *Kuetz.*
'Challenger'! *Ridley, Lea, and Ramage*!
Geogr. Distr. Atlantic.

PHÆOPHYCEÆ.

ASPEROCOCCUS INTRICATUS, *J. Ag.*
'Challenger'!
Geogr. Distr. West Indies.

DICTYOTA CILIATA, *J. Ag.*
'Challenger'!
Geogr. Distr. West Indies.

D. DICHOTOMA, *Huds.*
Ridley, Lea, and Ramage!
Geogr. Distr. Atlantic and New Zealand.

D. BARTYRESIANA, *Lam.*
'Challenger'!
Geogr. Distr. West Indies.

ZONARIA LOBATA, *Ag.*
Ridley, Lea, and Ramage!
Geogr. Distr. Atlantic (Cape of Good Hope to Canaries and Brazil).

PADINA PAVONIA, *L.*
'Challenger.' *Ridley, Lea, and Ramage*!
Geogr. Distr. Throughout warm seas.

HALYSERIS JUSTII, *Lamx.*
Ridley, Lea, and Ramage!
Geogr. Distr. West Indies.

H. DELICATULA, *Lamx.*
'Challenger'! *Ridley, Lea, and Ramage.*
Geogr. Distr. Atlantic (America from Mexico to Brazil).

H. PLAGIOGRAMMA, *Mont.*
'Challenger'!
Geogr. Distr. Atlantic (America from West Indies to Brazil) and Pacific (Sandwich Islands).

SARGASSUM VULGARE, *Ag.*
'Challenger'! *Ridley, Lea, and Ramage* !
Geogr. Distr. Tropical and South subtropical Atlantic.

CHLOROPHYCEÆ.

BRYOPSIS PENNATA, *Lamx.*
Ridley, Lea, and Ramage!
Geogr. Distr. West Indies (Indian Ocean?).

CODIUM TOMENTOSUM, *Ag.*
Ridley, Lea, and Ramage!
Geogr. Distr. Throughout tropical and temperate seas.

HALIMEDA OPUNTIA, *Lamx.*
'Challenger.'
Geogr. Distr. Throughout tropical seas.

VALONIA FILIFORMIS, *Dickie.*
Ridley, Lea, and Ramage!
Geogr. Distr. Mauritius.

V. VENTRICOSA, *J. Ag.*
Ridley, Lea, and Ramage !
Geogr. Distr. West Indies.

Webster, in 'Voyage to the Southern Ocean in H.M.S. 'Chanticleer,'' vol. ii. p. 337, gives a description of "clusters of vesicles" which he cannot determine as animal or vegetable. The description obviously applies to *Valonia ventricosa*, even to minute details. He describes only one other seaweed from Fernando Noronha. It had "the aspect of a land plant, leaves linear, pinnate, an inch in length, and of a very bright green." This would apply to either a *Caulerpa* or a *Bryopsis*.

CHAMÆDORIS ANNULATA, *Mont.*
Ridley, Lea, and Ramage!
Geogr. Distr. Tropical Atlantic and Indian Ocean.

CAULERPA PROLIFERA, *Lamx.*
Ridley, Lea, and Ramage!
Geogr. Distr. Tropical and subtropical Atlantic and Mediterranean.

C. TAXIFOLIA, *Ag.*
Ridley, Lea, and Ramage!
Geogr. Distr. West Indies (Pacific? Australia?).

C. CUPRESSOIDES, *Ag.*, var. ALTERNIFOLIA, *Crouan.*
Ridley, Lea, and Ramage!
Geogr Distr. West Indies.

C. CLAVIFERA, *Ag.*
'Challenger.' *Ridley, Lea, and Ramage!*
Geogr. Distr. Throughout tropical seas.

C. MEXICANA, *Sond.*
'Challenger.'
Geogr. Distr. Tropical Atlantic.

ENTEROMORPHA COMPRESSA, *Grev.*
Ridley, Lea, and Ramage!
Geogr. Distr. Cosmopolitan.

ENTEROMORPHA, sp.?
'Challenger.'
Geogr. Distr.

KALLONEMA OBSCURUM, *Dickie.*
'Challenger.'
Geogr. Distr. Known only from Fernando Noronha.

ULVA LACTUCA, L.
Ridley, Lea, and Ramage!
Geogr. Distr. Cosmopolitan.

U. LOBATA, *Kuetz.*
'Challenger.'
Geogr. Distr. Atlantic.

CLADOPHORA SUBVARICOSA, *Dickie.*
'Challenger'!
Geogr. Distr. Known only from Fernando Noronha.

C. MINUTA, *Dickie.*
'Challenger.'
Geogr. Distr. Known only from Fernando Noronha.

CLADOPHORA MORRISIÆ, *Harv.*
Ridley, Lea, and Ramage !
Geogr. Distr. North America.

CHÆTOMORPHA ANTENNINA, *Kuetz.*
Ridley, Lea, and Ramage !
Geogr. Distr. West Indies, South America, and India.

LYNGBYA NORONHÆ, *Dickie.*
' Challenger '!
Geogr. Distr. Known only from Fernando Noronha.

HORMOSPORA PELLUCIDA, *Dickie.*
' Challenger '!
Geogr. Distr. Known only from Fernando Noronha.

There are two imperfect specimens of *Cladophora* and several *Cyanophyceæ* in a condition which would not justify my assigning them names.

FUNGI.
By GEORGE M. MURRAY, F.L.S.
BASIDIOMYCETES.

LENZITES ERUBESCENS, *Berk.*
Ridley, Lea, and Ramage !
Geogr. Distr. Rio de Janeiro.

SCHIZOPHYLLUM COMMUNE, *Fr.*
Ridley, Lea, and Ramage ! Cape Placellière woods.
Geogr. Distr. Throughout the world, but never common.

POLYPORUS LUCIDUS, *Fr.*
Ridley, Lea, and Ramage ! On a stump in the Sapato.
Geogr. Distr. Cosmopolitan.

P. HIRSUTUS, *Fr.*
Ridley, Lea, and Ramage ! On a *Sapium* in the Sapato.
Geogr. Distr. Cosmopolitan.

P. FOMENTARIUS, *Fr.*
Ridley, Lea, and Ramage ! Common on Cape Placellière in one or two spots.
Geogr. Distr. Europe, Asia (Siberia and Penang), and North America.

EXIDIA GLANDULOSA, *Fr.*
Ridley, Lea, and Ramage ! Sapato woods.
Geogr. Distr. Europe, N. Asia, N. America, S. Africa, Australia, and Tasmania.

CYATHUS STRIATUS, *Hoffm.*
Ridley, Lea, and Ramage! Sapate woods.
Geogr. Distr. Europe, N. America, and Tropical Africa.

PYRENOMYCETES.

CLAVICEPS PURPUREA, *Tul.*
Ridley, Lea, and Ramage!
On *Cenchrus echinatus* in the coco-nut plantation, Sueste.
Geogr. Distr. Europe, North America, and N. Zealand.

DOLDINIA CONCENTRICA.
Ridley, Lea, and Ramage! Sapate woods.
Geogr. Distr. Cosmopolitan.

USTULINA VULGARIS, *Tul.*
Ridley, Lea, and Ramage !
Geogr. Distr. Europe, America, Ceylon.

XYLARIA POLYMORPHA, *Grev.*
In the Sapate woods on rotten trees.
Geogr. Distr. Whole world.

DIATOMACEÆ.

By JOHN RATTRAY, M.A., B.Sc.
(Communicated by H. N. RIDLEY, M.A., F.L.S.)

The following list of *Diatomaceæ* must be regarded as giving but a very imperfect conception of this department of the flora of Fernando and, indeed, of the collections made during the present Expedition. The materials entrusted to me for examination by Mr. H. N. Ridley consisted of :—a few small samples of calcareous Algæ, chiefly *Lithothamnion* and *Galaxaura*, in the dry state; similar but larger samples of the same genera preserved in spirit along with other *Chlorophyceæ* and *Rhodophyceæ*; a small sample of sediment from various gatherings, also in spirit; and dried specimens of *Ceramia* with shell-fragments. Two samples of guano from Rat Island off Fernando Noronha were also examined, and the species observed in these are given in a separate list. No surface-gatherings, which might be expected to yield many *Chætocerotidæ*, *Rhizosoleniæ*, and *Coscinodisci*, such as are abundant on the opposite shores of Africa and in the Gulf of Guinea, were taken.

Fam 1. CYMBELLEÆ.

1. AMPHORA MARINA, *W. Sm.*=A. lincolata, *Kuetz.*
Specimens similar to that in Schmidt's Atl. d. Diatk. pl. xxvi.
fig. 81, and approaching *A. acutiuscula*, Kuetz. Rare.

2. CYMBELLA OBTUSA, *Greg.* (?).
Girdle aspect only observed. Length ·03 mm.; breadth ·006 mm.
Markings 6 to 8 in ·01 mm. Rare.

3. C. AMPHICEPHALA, *Naeg.*
Length ·02 mm.; breadth ·006 mm. Rare.

4. COCCONEMA, sp. ?
Close to *C. australicum*, A. Schmidt. One fragment.

Fam. 2. NAVICULEÆ.

5. NAVICULA LACINIOSA, *A. Schmidt.*
Similar to Java specimens, but the extremities slightly more
acute. Length ·02 mm.; breadth at widest areas ·0075 mm. Few.

6. N. SUBULA, *Kuetz.*, var. *Grun.* (Ver. Wien. Ak. 1860, p. 548,
pl. i. fig. 24).
Length ·0375 mm.; breadth ·005 mm. Rare.

7. N. COCCONEIFORMIS, *Kuetz.*
Length ·015 mm.; breadth ·0075 mm. Rare.

8. N. MUTICA, *Kuetz.*
Length ·015 mm.; breadth ·0075 mm. Few.

9. N. WEISSFLOGII, *A. Schmidt.*
Similar to specimens from Sandwich Islands. Length ·0175
mm.; breadth ·0075 mm. Rare.

10. N. sp.?
Length ·0875 mm.; breadth ·0085 mm. Markings not visible
owing to the presence of detritus. One specimen.

11. N. BRASILIENSIS, *Grun.*
Length ·025 mm.; breadth ·0175 mm. Few.

12. N. MINUSCULA, var. BAHUSIENSIS, *Grun.*
Length ·025 mm. Markings obscure. Fragmentary. One
specimen.

13. N. ERYTHRÆA, *Grun.*
Length ·08 mm.; breadth at middle ·0325 mm. Striæ 8 in
·01 mm. Rare.

14. NAVICULA INTERRUPTA, *Kuetz.*, var.

Length ·08 mm.; breadth of central portion ·02 mm., of lobes ·03 mm. Rare.

15. PLEUROSIGMA SPECIOSUM, *W. Sm.*, var.

Similar to Tahiti specimens procured by H.M.S. 'Challenger.' Length ·1225 mm.; breadth ·025 mm. Several.

16. P. LORENZII, *Grun.*

Similar to specimens found by Dr. Lorenz in 2 to 4 fathoms on *Zostera*-ground in the Adriatic. Length ·1 mm.; breadth ·02 mm. Few.

Fam. 3. ACHNANTHEÆ.

17. ACHNANTHES SUBSESSILIS, *Ehrenb.*

Length ·015 mm. Rare.

18. A. GLABRATA, *Grun.*

Typical. Rare.

Fam. COCCONEIDEÆ.

19. COCCONEIS SCUTELLUM, *Ehrenb.*

Major axis ·02 mm., 1¾ times minor. Not uncommon.

Fam. 4. FRAGILARIEÆ.

20. SYNEDRA AFFINIS, var. HYBRIDA, *Grun.*

Length ·1025 mm.; breadth ·0125 mm. Striæ 12 to 13 in ·01 mm. This var. is allied to *S. tabulata*, Kuetz., and *S. nitzschioides*, Grun.

21. S. AFFINIS, var. DELICATULA, *Grun.*

Length ·0625 mm.; breadth ·005 mm. Specimens found attached laterally. Common.

22. S. OXYRHYNCHUS, *Kuetz.*

Length ·0625 mm.; breadth ·006 mm. Rare.

23. S. ACUS, *Kuetz.*

Length ·0625 mm.; breadth ·005 mm. Rare.

24. S. LANCEOLATA, *Kuetz.*

Length ·075 mm.; breadth ·01 mm. Few.

25. LICMOPHORA AUSTRALIS, *Grun.*, var. MAJOR.

Length ·0425 mm.; breadth ·025 mm. Rare.

26. L. DEBILIS, *Grun.*=Podosphenia debilis, *Kuetz.*

Length ·0225 mm.; breadth ·0125 mm. Few.

27. L. PARADOXA, *Ag.*=Diatoma flabellatum, *Juerg.*, et Gomphonema paradoxum, *C. Ag.*

Length ·04 mm.; breadth ·035 mm. Few.

28. LICMOPHORA LYNGBYI, var. LONGA, *Grun.*
Length ·2625 mm.; breadth ·05 mm. One specimen observed.

Fam. 5. TABELLARIEÆ.

29. GRAMMATOPHORA MARINA, var. INTERMEDIA, *Grun.*
Concatenate. Common.

30. G. ANGULOSA, var. HAMULIFERA, *Grun.*
Similar to specimens from Valparaiso. Length ·015 mm.; breadth ·0175 mm. Several.

31. G. JAPONICA, *Grun.*
Length ·065 mm.; breadth ·0125 mm. Concatenate. Rare.

Fam. 6. SURIRELLEÆ.

32. NITZSCHIA MARGINULATA, var. SUBCONSTRICTA, *Grun.*
Length ·0625 mm.; breadth at median constriction ·0175 mm., at lobes ·02 mm. Rare.

33. N. LANCEOLATA, *W. Sm.*
One fragment.

34. N. MARINA, *Grun.*
Length ·13 mm.; breadth ·0125 mm. Few.

35. N. FLUMINENSIS, *Grun.*
Similar to Campeachy Bay specimens. Length ·12 mm.; breadth ·01 mm. Rare.

Fam. 7. CHÆTOCEROTIDÆ.

36. RHIZOSOLENIA STYLIFORMIS, *Brightw.*
Calyptriform extremity only observed. One specimen.

Fam. 8. MELOSIREÆ.

37. MELOSIRA NUMMULOIDES, *Ag.*
Concatenate. Abundant.

Fam. 9. BIDDULPHIEÆ.

38. BIDDULPHIA PULCHELLA, *Gray.*
Concatenate. Common. In the same chain a frustule ·125 mm. long had not divided, whilst the second from it, of the same length, showed the division complete. The breadth of the girdle before division is about ·0625 mm. Abundant.

39. B. MOBILIENSIS, *Grun.* = B. Baileyi, *W. Sm.*
Common on the Rhodophyceæ. Not associated with the foregoing.

40. TRICERATIUM FAVUS, *Ehrenb.*
One fragment observed on *Ceramium rubrum.*

41. T. ELEGANS, *Grev.*=T. Hardmanianum, *Witt.*
Diam. ·04 mm. Rare.

42. T. ALTERNANS, *Ehrenb.*
3-sided, each side with two equal concavities. Diam. ·03 mm.
One specimen.

43. T. PENTACRINUS, *T. Wallich.*
4-sided var. Diam. ·0675 mm. One specimen.

Fam. 10. COSCINODISCEÆ.

44. COSCINODISCUS ANGUSTE-LINEATUS, *A. Schmidt.*
Diam. ·0325 mm. Markings 6 in ·01 mm. Few.

45. C. DENARIUS, *A. Schmidt.*
Diam. ·0875 mm. Several.

46. C. MINOR, *Ehrenb.*
Diam. ·045 mm. Rare.

GUANO SPECIMENS.

Fam. 1. BIDDULPHIEÆ.

1. TRICERATIUM FAVUS, *Ehrenb.*
Side ·1625 mm. long. Rare.

2. T. TRISULCUM, *Bailey.*
Side about ·25 mm. long. One specimen.

Fam. 2. EUPODISCEÆ.

3. AULISCUS CÆLATUS, var. STRIGILLATA, *A. Schmidt.*
Major axis ·09 mm., minor ·0825 mm. One specimen.

Fam. 3. HELIOPELTEÆ.

4. ACTINOPTYCHUS SPLENDENS, *Ralfs.*
Diam. ·12 mm. One specimen.

Fam. 4. COSCINODISCEÆ.

5. ARACHNOIDISCUS EHRENBERGII, var. CALIFORNICA, *A. Schmidt.*
Diam. ·24 mm. One specimen.

6. COSCINODISCUS BIANGULATUS, *A. Schmidt.*
Diam. ·15 mm. Several.

7. Coscinodiscus heteroporus, *Ehrenb.*

Diam. ·14 mm. A very small central space, the markings largest around the centre and at about ¾ of the radius from it, beyond this diminishing on a somewhat wide zone to the lower.

8. C. marginatus, *Ehrenb.*
Diam. ·095 mm. Several.

9. C. perforatus, *Ehrenb.*
Diam. ·125 mm. Markings increasing slightly outwards to 3 in ·01 mm. Small on a narrow band within the border.

10. C. intumescens, *Pant.* (?).
Diam. ·0875 mm. Rare.

11. C. oculus-iridis, *Ehrenb.*
Diam. ·2 mm. Few.

12. C. argus, *Ehrenb.*
Diam. ·1125 mm. Rare.

13. C. radiatus, var. media, *Grun.*
Diam. ·1075 mm. Few.

14. C. robusta, *Grev.*
Diam. ·0875 mm. Rare.

15. C. asteromphalus, *Ehrenb.*
Diam. ·25 mm. Rare.

GEOLOGY,

Based on Petrological Notes by Thomas Davies, F.G.S.

(Communicated by H. N. Ridley, M.A., F.L.S.)

The whole cluster of islands now above water only presents two groups of igneous rocks, viz. phonolites and basalts, to which must be added, by way of sedimentary rocks, some raised coral-reefs and some sandstone formed of blown sand.

The phonolite is confined to Sella Giueta, the Peak, Look-out Hill, Tangle rock, Morro branco, the Central Knoll, and the island known as the Frade; but at Sponge Bay, on the south-east side of the Main Island, are some beds of phonolitic tuffs traversed by dykes.

It is usually rudely columnar, with the columns generally forming an angle with the horizon; but on the Central Knoll they are erect. Near Tangle Rock are horizontal columns so regular that Mr. Ramage found in one spot a complete tunnel through the mass formed by one or two having fallen out.

Morro branco ("the white hill") is a dome-shaped hill of a white rock, very different in appearance to the normal phonolite and somewhat resembling some specimens of the tuffs near Sponge Bay. It appears to have been altered by contact with basalt. A boulder of similar rock was cut through at Leao Bay by the convicts, who were making a road there. On Morro branco were found several plants which grew only here and on the similar rock of Look-out Hill, and one species of grass was peculiar to the former spot.

Not only at Morro branco but on the western cliffs beyond the peak at Boldro we found phonolite rocks altered by contact with basalt, evidencing the fact that the phonolite was anterior to the basalt. It seems, in fact, to form a groundwork of the island, covered in part with later deposits of basalt, through which project the high hills and peaks.

In Sella Gineta and Tangle Rock were strata of a muddy chert passing into a form of semiopal. This appears to be the residue of hot siliceous springs, and must have been deposited between the pouring out of one layer of phonolite and the next. In one specimen is the mould of a very large crystal, apparently of calcite. From this rock, which appears to have been laid down horizontally, it may be gathered that the phonolite was not originally injected into older rocks and then elevated above the surface, the older rocks being weathered away, but rather that it was poured out at intervals between which the hot springs deposited their silica on the cold or cooling phonolites, which was afterwards covered with another layer of the phonolites.

Basaltic Rocks.

The larger part of the islands now consists of these rocks, which occur in the forms of columnar or spheroidal basalt, in layers or dykes, scorias, pumice, tuffs, volcanic agglomerates. In Rat Island the basalt rises from sea-level at the western end, where it is capped with raised reef, to high vertical cliffs; though chiefly consisting of the ordinary fine-grained olivine basalt of the island, here and there, at the isthmus which connects the eastern promontory with the main part of the island, vesicular and amygdaloidal

basalt tuffs and scoriaceous rock from the top of a lava-flow
were obtained.   Passing west from Rat Island the basalt is almost
entirely submerged till Platform Island (San José).   This island
is about 76 feet in height and crowned with reef-rock and some
indurated sand-rock resembling quartzite.   There are two smaller
detached rocks of columnar basalt on the N.W. side.

San José is connected with the mainland by a low ridge covered
with pebbles, bare at very low tides, the remains apparently of a
peninsula of which Morro do Chapeo, a small rock about 12 feet
in height capped with raised reef, is all that is left above sea-level.
Between this rock and the Main Island were a number of large
boulders of very hard basalt, containing "bombs" or balls of
olivine, enstatite, and augite, much resembling those of the
Laacher See, and very fresh in appearance.

In the Main Island the basalt appears in the western extremity
or almost at sea-level and rises gradually towards the east hills,
and after forming the great plateau of the Central district passes
away into Cape Placellière and Point Noir, in the western and
north-western ends of the isle.   It occurs in the form of horizontal layers at Tobacco Point, and round the lake, at Boldro and
elsewhere, and in some spots, especially at the last-mentioned place,
it was not difficult to find the bands of scoriaceous and pumiceous
rock which had formerly formed the upper layer of a lava-flow,
which had again been covered by another and another lava-deposit.
The basalt here, again, was rich in olivine and augite, and the
prevailing rock was fine-grained, black, and compact.   Besides
these layers were other masses, columnar in structure, the columns
being usually larger than those of the phonolite.   They were very
well seen in Portuguese Bay.   At other spots, as at Sponge Bay,
East Hills, and Tangle Rock, the basalt was in spheroids, sometimes of immense size.   Near Tangle Rock they were much
decomposed and altered, and showed the presence of much iron.

Dykes.—Dykes of basalt occurred in several parts of the island.
At Sponge Bay they are numerous, running down from the East
Hills into the sea ; here they traverse beds of phonolitic tuffs, and
some are as much as three feet in thickness ; they are often
transversely columnar, and curved and even branched.   Similar
dykes occur in Leao Bay, here running north, and on the north
side of the island, near Cape Placellière, where they run north-east,
traversing beds of scoria.   They are here of considerable height,

sometimes, in fact, the whole height of the cliff, but are not very broad.

Scorias, Pumices, and Tuffs.

At the eastern end of the island is a large quantity of a red clayey soil, covered in part by some sand-dunes, and apparently overlying the ordinary basalt. This appears to be a scoriaceous basalt which has been much altered by the action of acid vapours. Another large band, harder in texture, runs from the low hills on the south of the road from San Antonio to Chaloupe Bay, where it crops out between two masses of basaltic rocks ; it is of considerable depth and 100 yards in width. The central plateau of the island is covered also with a somewhat similar red clay-like deposit, which is very fertile and thickly covered with fodder-plants &c. This appears to be a ferruginous scoriaceous rock, probably poured out from a volcano in the form it is at present, and not the product of decomposition of basalt, as it at first appears.

At Look-out Hill, Boldro, and other places where the basalt comes in contact with the phonolite and has not been much compressed, it was loose in texture and frequently amygdaloidal, with zeolites.

Between Morro branco and Point Noir was a very interesting series of volcanic rocks. After passing the lake, which is surrounded by a high semicircle of cliffs of basalt arranged in layers, we come to a promontory consisting of an immense barren mass of large fragments of basalt, irregular in size and shape and piled on the top of each other to a great height above the sea. They appear to be broken columns of fine-grained basalt, and are known as Pedras Pretas (the Black Rocks). Beyond these is an indentation terminated at the west by Cape Noir, a black cape of basalt. Between the Cape and Pedras Pretas is a steep slope, about 700 feet high, consisting of a thick bed of scoria, in the shape of rolled balls overlain by a thick bed of basalt in layers. The scoria-bed was not flat, but was steeply sloped from the top of the hill to the level of the sea, the basalt following the curve. Here, we had no doubt, was a bed of ash and scoria thrown out at the first eruption of a volcano at no great distance from Cape Placellière, which still retained the slope at which they eventually settled when falling from the crater and were then covered by the flows of lava which followed this first eruption, and which has preserved the ash-beds from destruction and denudation to the present day. The centre of the promontory which ends in Cape Placellière consists at the top of a bluish, rather loose-textured basalt; but the dense thickets and woods here make the geology difficult to see. On the north side of the Sapate, however, we come again to scoria-beds in the cliffs traversed, as above mentioned, by basalt dykes; and at Cape Placellière itself we found scoriaceous rocks, with volcanic agglomerates, pumices, and other loose-textured lavas, which seem to confirm the theory of an important crater having formerly existed in the neighbourhood. The Cape itself, however, appears to be basalt, and we were unable to trace any remains of a crater in this direction. One can only suggest from the angle at which the bed of scoria above mentioned lies that the crater from which they were ejected was somewhere to the N.E. of Cape Placellière.

Mr. Ridley writes:—" In one spot on this side of the island we found a cavern eaten out by the sea; its walls were formed by two dykes traversing the scoria-beds; the softer scoria had been removed by the sea up to a certain height, forming a small cave; I mention this because I believe it to be the origin of the Hole in the Wall that pierces Cape Placellière, which was some distance beyond this. We were unable to reach the Hole because of the great difficulty of getting there by land, though we made several attempts; nor were we able to reach it by sea on account of having no boats. Cape Placellière consists of a narrow high wall of basalt, scoria, and pumiceous and tufaceous rocks, running almost due north, the base of which is of some thickness, and the top a narrow broken edge. As far as we could make out from the nearest point we could reach, a dyke runs half across the entrance to the Hole on the eastern side.

"Sand occurs in several of the larger bays, chiefly on the north side; but in Peak Bay we were informed that during the winter months, when the current sets this way from the north, all this sand is removed, and the large basalt boulders underlying it are uncovered. At certain spots and at certain times black sand is found on the shore; this seems to be produced by the washing-out of the lighter grains of quartz and fragments of shells &c., so that only the heavier hornblende and magnetite crystals and grains are left."

Above S. Antonio Bay are extensive sand-hills, the sand being drifted up by the wind from the south. In some spots it was cemented together by carbonate of lime from decomposition of shells, and on the top of San José we found some masses of a hard quartzite-like sandstone, apparently formed of blown sand and including a *Corbula* and a *Venus*, neither species met with elsewhere. In this rock also was a pseudomorph of apparently a felspar crystal, about an inch long, consisting of pure white translucent quartz. There was not much of this rock, and it is quite possible that it might not belong to the island.

Salt crystallizes out here and there on San José Island and elsewhere, from evaporation of sea-water; and calcite coats the raised coral-reef at Tobacco Point, and perhaps is the cementing material which in some spots of Peak Bay forms the pebbles of the beach into a conglomerate. It also lines and fills up cracks and fissures in the softer basaltic rocks at Cotton-tree Bay and in the phonolitic tuffs at Sponge Bay.

General Summary.

From these Notes it will be seen that the islands are, as has been constantly affirmed, of volcanic origin, and further that we can trace two distinct periods in their history, the phonolitic and the basaltic periods; that the phonolite was ejected in the form of phonolitic lavas and tuffs, and that there were periods of cessation of action between the eruptions, during which some hot spring deposited beds of silica. After this had happened, and perhaps at a much later date, and after much denudation had taken place, craters in the north-west and south-east portions of the island ejected scorias, pumices, tuffs, and basalt, which covered a great portion of the phonolitic rocks and altered them where it came in contact with them. At a later period, probably when volcanic action had ceased, portions of the submerged basalt beds became covered with a thick deposit of coral-reef, sometimes 100 feet in thickness, which was perhaps gradually raised above sea-level to a height of from 3 to 4 to 100 feet, and which, in some spots, having been used by the sea-birds for many years as a resting- and probably as breeding-places, became covered with a deep layer of guano, on which a rich vegetation soon established itself. All volcanic action seems to have ceased for many years, there are no traditions of tidal waves or earthquakes, and the early discoverers, nearly 400 years ago, noticed no signs of volcanic activity, such as were visible at that time in the Canary Islands. Much denudation and destruction has without doubt occurred, but the soundings which have been taken round the island are insufficient to give us any clear idea of its original size. At a short distance on each side of the group, the depth suddenly increases to over 2000 feet.

Some American petrologists, who have found similar rocks to those of Fernando do Noronha in the neighbourhood of Cape San Roque, seem to consider that the group may have been connected at one time with the mainland at this point. We have not seen the specimens, but from the form and arrangement of the rocks here it may be doubted that the evidence is sufficient to prove a connection, while the presence and position of the scoria-beds of Pedras Pretas seem most clearly to prove that there was here a large and active-enough crater to supply nearly all the basalt upon the island.

Now all volcanic activity has long ceased and the last stage in the geological history of the island, that of its breaking up into smaller islets and its slow destruction by denudation, has been reached.

The Coral Reef.

The reef lying round the group of islands, though of considerable extent, does not entirely surround them, but occurs in irregular patches of varying extent. As much interest has been lately shown in the structure and origin of reefs, a few notes made here may be of some value. The reef, in structure, is a whitish-brown, friable, calcareous rock, which, when broken, shows, except for a short way below the surface, no identifiable animal- or plant-remains. It weathers into caves and hollows and rock-pools, so that there are often caverns excavated deep underneath the ledges, into which, at low water, the sea rushes with great velocity expelling the air violently.

The largest of these blow-holes is at Rat Island, where the rush of spray through the hole attains so great a height that it can be seen for some miles.

Where the reef has been raised above sea-level and subjected to aerial decomposition, it seems to be oolitic in structure and finely granular. In comparing it with the reef-rock of the well-known Recife at Pernambuco, it seems that the latter is harder and more compact.

In the rock-pools and crevices of the reef live an immense abundance of marine animals—corals, sponges, echinoderms, worms, crustacea, and mollusca being very abundant; and over and round the edges grow many calcareous Algæ and Foraminifera. Although it is natural to talk of it as coral-reef, corals contribute but a small share of its structure. A broken piece of reef shows that layers of *Lithothamnion*, and other calcareous Algæ, with the tubes of *Serpulæ, Polytrema rubrum*, and calcareous *débris* form the greater part of the mass. These, however, are only distinguishable at the top of the living reef. About an inch ·below the surface the remains become indistinguishable.

With respect to the distribution of the reef, it may be noted that it apparently never forms where the cliffs descend directly into the sea, nor on the shores covered with large loose boulders, nor in the sandy bays, but it seems to me that its formation here is to a certain extent dependent on the streams which pour into the sea at different points. Thus, there are streams at almost if not all the living reefs, viz., at Chaloupe Bay, Sponge Bay, and Sambaquichaba.

The Recife, too, at Pernambuco is at the mouth of the river, and at Itamaraca, further along the coast to the north, Mr. Ramage, who visited it, reports very extensive reef; and here, again, rivers enter into the sea.

I imagine that the nullipores, corals, and other plants and animals which make the reef cannot grow upon sand which is always shifting, nor upon irregular boulders; but where the sand becomes mixed a little with silt, or the gravel consolidated by it, they can grow and thrive.

The reef grows only in water just below high-water mark, and abruptly terminates in ledges, beneath which are usually hollows and caverns. The chief growth is along the edges exposed to the waves. Sometimes, as at Sponge Bay, the outer edge of the reef is much higher than the inner portion, probably owing to the more rapid growth here of the nullipores.

The shore of the neighbouring bays consists of sand or basalt pebbles, and there is no more reef till Sambaquichaba, where a narrow spit or two runs out into the sea. We saw no more reef on this side of the island. On the south side, beginning again at the east, there is a very extensive reef in Sponge Bay, of considerable thickness and covered at high water. Beyond the reef, exposed at low water, there seems to be a lower ledge of great extent. Far out at low water can be seen two rocks just raised above high water, over which the sea constantly breaks. These are, I believe, the rocks marked on the Admiralty Chart as "the Brothers." The name is not known to the inhabitants, who have given the same appellation to the twin rocks off Sambaquichaba. Passing round the coast, we come next to Cotton-tree Bay, where there is a reef some little way from the shore and covered at high water. On the west corner of the bay is a deposit of reef-rock, 100 feet above sea-level and 100 feet in thickness, overlying basalt. No more is met with till Tobacco Point, where is a large deposit of raised reef.

The living reef at the present day occurs on the islands in a number of spots, not continuously, but here and there, sometimes in the form of short spits, at others covering large extents of the sea-bottom. In Rat Island it covers the whole of the western corner and attains considerable thickness in parts ; but the only living reef is on the south-west angle, where the sea beats violently from the south. The rest of the corner consists of a deposit of dead reef, with a layer of guano from 4 to 10 metres thick overlying it. In one spot it overlies a beach of basalt-pebbles of large size, which is continuous with a similar uncovered beach to the north. The dead reef projects in the form of weathered pinnacles all over this part of Rat Island and also in Booby Island, which is evidently a continuation of the same reef broken through after being raised above sea-level by the waves. The reef here rings to the hammer, and is very hard and compact ; it appears to have much sand in it. Egg Island is apparently also a continuation of this reef.

The island known as San José, or Platform Island, is composed of a basis of basalt, still connected with the mainland by a band of basalt, forming a kind of ridge of pebbles, bare at very low tides. It appears to be the remains of a large promontory, of which the Morro do Chapeo forms a part. It is a basalt rock about 90 feet high, is capped with reef about 6 feet thick, and containing more distinguishable animal-remains than most of the raised reef. The reef on Morro do Chapeo is about the same height above sea-level (about 10 feet) as that on Egg Island, and very much lower than that on San José. Passing along the north side of the Main Island, there is no reef till we reach Chaloupe Bay, the shore consisting of large basalt pebbles, with very large crystals of olivine and enstatite at the extreme eastern point, and sand from San Antonio Bay to Chaloupe Bay. In Chaloupe Bay there is a large patch of living coral-reef, extending the whole length of the bay, and entirely covered at high water.

EXPLANATION OF THE PLATES.

PLATE I.

Erythrina aurantiaca, Ridl.

Fig. 1. Half-expanded flower, side view. Figs. 2, 3. Alæ. Fig. 4. Andrœcium and pistil. Fig. 5. Pistil. Fig. 6. Seed.
(Figs. 3–5 magnified.)

PLATE II.

Cyperus circinatus, Ridl.

Fig. 1. Entire plant. Fig. 2. Flower, magnified.

Oxalis sylvicola, Ridl.

Fig. 3. Entire plant. Fig. 4. Stamens and pistil, magnified.

PLATE III.

Sapium sceratum, Ridl.

Fig. 1. Flower-buds. Fig. 2. Inflorescence. Figs. 3 and 4. ♂ flowers. Fig. 5. Stamen. Figs. 6 and 7. ♀ flowers.
(All the details enlarged.)

PLATE IV.

Paspalum phonoliticum, Ridl.

Fig. 1. Diagram of flower. Figs. 2 and 3. Flowers, the details enlarged.

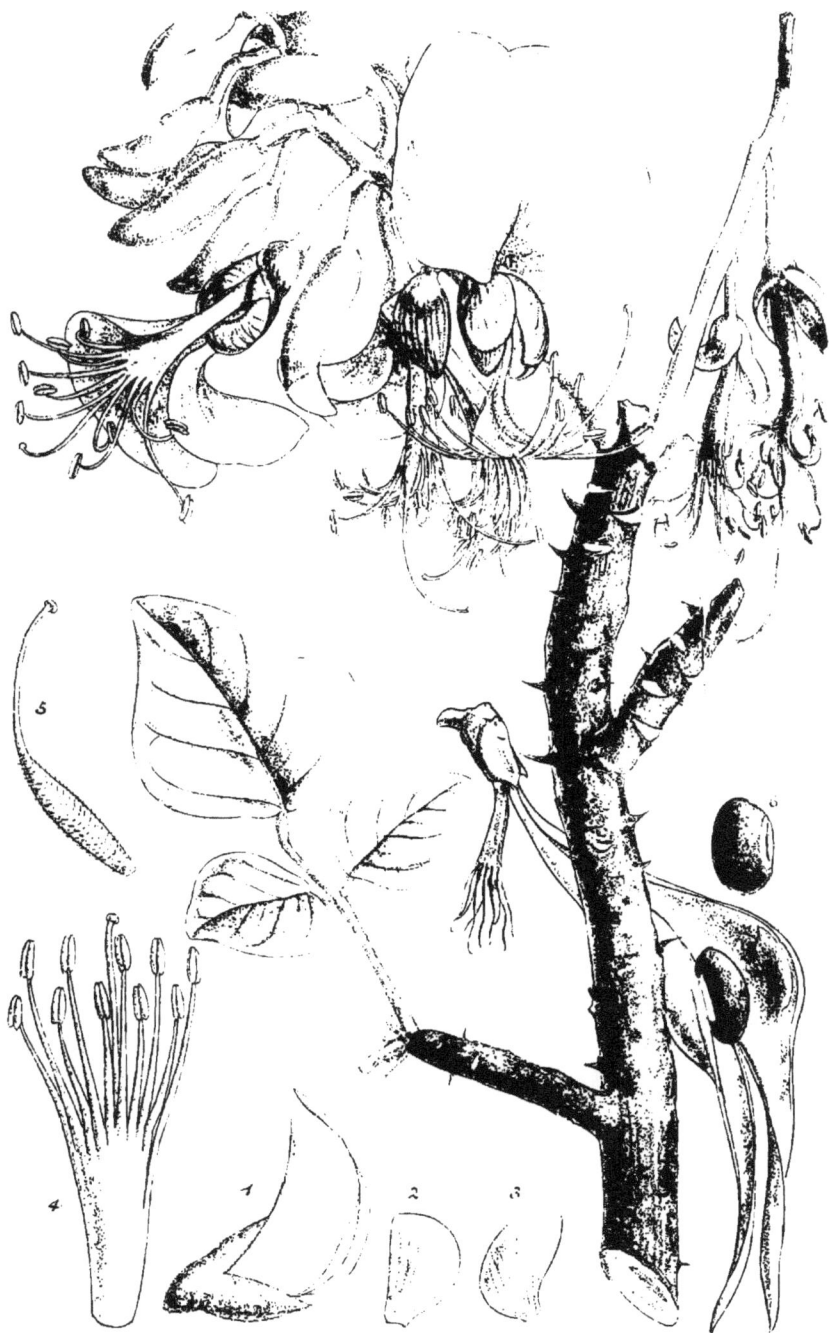

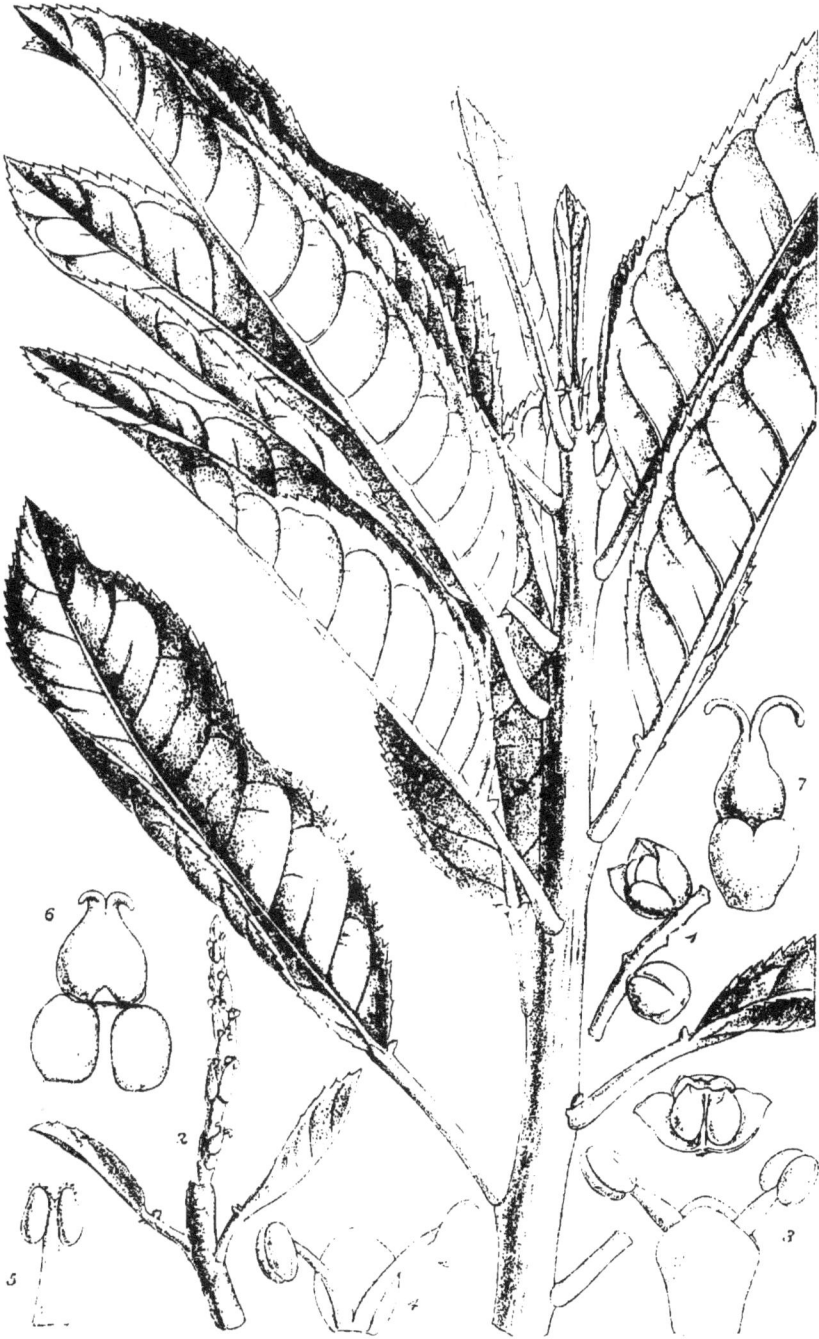

PASPALUM PHONOLITICUM

NOTES ON THE ZOOLOGY OF FERNANDO NORONHA.
By H. N. RIDLEY, M.A., F.L.S.

[Read 7th June, 1888.]

(PLATE XXX.)

INTRODUCTION.

On July 9th, 1887, the writer, with Mr. G. A. Ramage, of Edinburgh, started for Brazil to thoroughly explore the island of Fernando Noronha, lying in long. 32° 25′ 30″ W. and lat. 3° 50′ 10″ S., at a distance of 194 miles N.E. from Cape San Roque, coast of Brazil. On arriving at Pernambuco we were joined by the Rev. T. S. Lea, who came as a volunteer at his own expense. The cost of the expedition was defrayed by the Royal Society. After some delay at Pernambuco we embarked in the 'Nasmyth' steamship, trading to Liverpool, which was permitted to land us at the island, as the regular steamer trading between Pernambuco and Fernando Noronha was detained for a long time just as she was due to start. We arrived at our destination on August 14th, and remained there till September 24th, when we returned by the little Brazilian steamer to the mainland. We occupied ourselves in exploring, and in collecting plants, animals, and rock-specimens in all parts of the main islands, and visited also most of the other islets which were accessible; but owing to the absence of boats, which, on account of the convict-station, are not permitted on the island, we were unable to obtain much by dredging. The coral-reefs, however, at low tide afforded an abundant harvest of marine animals and plants.

Having in the 'Introduction' to my "Notes on the Botany of Fernando Noronha," printed in the 'Journal of the Linnean Society' (Botany, vol. xxvii. p. 1), given a detailed account of the group of islands of which this is the chief, as well as a history of its discovery by Amerigo Vespucci in 1503, it will be unnecessary to repeat what has there been stated. For the better understanding, however, of the special reports on Zoology which are now furnished, the following extracts from the Introduction referred to may be found useful.

Vespucci's description of the trees and innumerable birds is evidently correct, though most of the trees are destroyed, and the birds far less abundant than they were then. The lizards with two tails may have been a confusion of the very abundant and conspicuous Gecko with the *Amphisbæna*, which is often called the snake with two heads, or may have been suggested by finding an accidentally fork-tailed lizard, of which an example was obtained by our expedition. The "serpents" were doubtless the *Amphisbæna*. The large rats are much less easy to explain ; at present the only rats occurring on the island are *Mus rattus*, the common introduced black rat. It is impossible that the animals seen by Vespucci could have been this species, which could not at that time have been introduced. It is possible that there was formerly an indigenous rat-like mammal, which became exterminated by the black rat. We could find no tradition even of this big rat, and I fear it is quite extinct. The only hope of recovering its remains lies in an examination of the guano deposits of Rat Island, where its bones might be preserved.

The number of insects belonging to the orders which are well known as plant-fertilizers is surprisingly limited. A few small species of moths haunted at night the bushes of *Scoparia dulcis*, *Cassias*, &c. on the open spaces. A single species of butterfly was very abundant on Rat Island and the main island, but we never saw it visiting flowers.

The most important fertilizer was a small endemic hornet belonging to the genus *Polistes*, which gathered honey from the Leguminosæ and Cucurbitaceæ ; and three small black species of *Halictus* were caught in the flowers of the melons, *Momordica charantia*, *Oxalis Noronhæ*, and the mustard. The last plant was also haunted by *Temnoceras vesiculosus*, a pollen-eating Syrphid. The only other insects which could also be considered as possible fertilizers were *Tachytes inconspicuus*, n. sp., and *Monedula signata*, two sand-wasps, *Pompilus nesophila*, n. sp. (Hymenoptera), and *Psilopus metallifer* (a Dipteron), but none of these were seen at or near flowers. A small black beetle also was found in the flowers of an *Acacia* in the Governor's garden.

Though the number of species of insects was not large, the individuals, especially of the *Polistes* and *Halicti*, were very numerous, but at the same time they seemed out of all proportion to the immense number of flowers to be fertilized. It is very probable, however, that the majority of the Leguminosæ and some of the other plants were self-fertilized.

The lake on the main island contained a species of *Nitella* and an alga, an aquatic beetle and an Hemipteron, a new species of

Planorbis, and an Ostracod, the latter also occurring in all the streams of any size. The remaining streams and puddles produced dragonflies, a species of *Gammarus*, and a few algæ. One may compare this state of things with the freshwater fauna and flora of the other Atlantic islands. The absence of freshwater fish and amphibians is common to most small islands.

Just as with plants, a considerable number of animals have been introduced by man into the islands intentionally and by accident : such, for instance, as the Gecko (*Hemidactylus mabouia*), the American Cockroach (*Blatta americana*), and its curious parasite *Evania*, a spider, centipede, scorpion, rats and mice, and *Sitoplidus oryzæ*. These, though usually plentiful on the main island around the houses, are markedly absent from the smaller islets.

There are also many visitors which have arrived here by the aid of their wings, probably assisted by a suitable wind. They include a number of the peculiar terrestrial fauna, the land-birds and the insects. On looking over the lists of species taken here, we may note that the smaller birds are endemic, and a large proportion of the smaller insects. The small butterfly and almost all the moths are known from the mainland of South America, and the dragonflies are also widely distributed forms. All the winged fauna have a South-American facies, whether they are endemic or of wider distribution.

There are other creatures unprovided with means of traversing the ocean and not introduced by man. They include the *Amphisbæna*, Skink, the freshwater and terrestrial Mollusca, and perhaps some of the feebler-winged and apterous insects, the endemic ostracod, &c.

The *Planorbis*, *Gammarus*, and Ostracod, all supposed to be endemic species, may possibly have been brought over on the feet of Wading birds, which migrate here.

The presence of some others is more difficult to account for. The Mollusca are almost all peculiar, and the two that are not so are West-Indian. The *Amphisbæna* and Skink are endemic, and allied not to Brazilian but to West-Indian forms.

It is commonly said that reptiles and terrestrial mollusks find their way across the ocean by secreting themselves, or their eggs, on floating trees, which are drifted to islands; and though for several reasons this does not seem a satisfactory explanation of their distribution, yet the appearance of these animals here suggests this as the means by which they may have arrived. As

I have said, they are West-Indian in their affinities, and it is a striking fact that the marine fauna and flora are mainly West-Indian, while at least one of the plants (*Ipomœa Tuba*) whose seeds are known to be constantly drifted about at sea, and thus carried from place to place, is also only known from the West Indies. Another fact of interest in connection with this sea-travelling fauna, if I may use the expression, is that almost all the species noted occur on all the islands suitable for their existence. Thus, on Rat Island the *Bulimus Ridleyi*, the *Amphisbœna*, and Skink are common on St. Michael's Mount; the Skink is a large species, but the island, being a mere rocky peak, is unsuited for the *Amphisbœna*.

On Platform Island the lizard and several terrestrial Mollusca were found, while at the same time almost all the animals of more recent introduction were absent from these localities, just as is the case in the distribution of the plants. I believe, in fact, that this part of the fauna and flora was established on the island before it was broken up into the little archipelago of rocks and islets of which Fernando Noronha now consists *. Perhaps even this portion of the fauna and flora was introduced previously to the deposition of the basalt over the masses of phonolite which form as it were the skeleton outline of the island.

MAMMALIA.

No indigenous Mammals are to be found on these islands, and notwithstanding their proximity to the mainland, where Bats are abundant, no Bat of any species was observed by us, nor had the convicts ever seen any. Rats and Mice are exceedingly common. The Rat (*Mus rattus*) is here much paler than usual, and generally of a grey colour, while albinos are sometimes met with. It frequents the melon-fields and the tops of the cocoanut-trees, and is very destructive. The common House-Mouse, *M. musculus*, is even more abundant, and has suggested the name Rat Island (Ilha do Ratta), where it is

---

* On reference to A. Vespucci's description of the place, it will be found that he speaks of it as one island, so the breaking-up into an archipelago can only have taken place within the last 400 years.

as common as on the mainland. It swarms everywhere, and is so tame that it is often caught by the hand. I have seen one in the evening on the top of the inflorescence of a *Crotalaria*, apparently devouring the young seed-pods. Albinos are often seen. There being no birds or beasts of prey to keep these animals in check, and food being particularly abundant, they have increased enormously, and one of the employments of a convict is to capture a certain number of rats and mice once a month. At the monthly rat-hunt while we were on the island over 3900 were taken ; but we were assured that, in the dry season, when the herbage which covered the greater part of the island was dried up and burnt, the mice were compelled to leave their holes, and many more were taken. The hunts are then undertaken weekly, and 20,000 have been caught in a day. The bodies are piled up in the square after evening service, and the numbers counted.

The Cat is said to have become feral on the main island ; and on Rat Island and one or two of the other islands we saw a large black Cat which had escaped from an Italian vessel wrecked there, and which had run wild.

In Amerigo Vespucci's account of the island above quoted, he mentions " *Mures quam maximi.*" What these were we cannot now determine, but it is highly improbable that they were *Mus rattus*.

A species of Dolphin was constantly seen in San Antonio Bay and also off Rat Island. One was captured during our visit : its stomach contained many cuttlefish and prawns, the latter very similar to the common edible prawn of Pernambuco. Whales also passed within sight of the island on one occasion, but we did not see them.

AVES.

By R. BOWDLER SHARPE, F.L.S., &c.,
Assistant in the Zoological Department, British Museum.

The birds of the island are not very numerous as regards species, and apparently there are only three indigenous Land-birds. The species of Sea-birds found by Mr. Ridley are precisely what one might have looked for, but it is a little remarkable that no Petrel was observed.

Fam. VIREONIDÆ.

1. VIREO GRACILIROSTRIS, sp. n.

V. similis *V. magistro,* et forsan proximus, sed forma graciliore, coloribus dilutioribus, facie laterali pallide flavicante, et rostro valde tenuiore et graciliore distinguendus. Long. tot. 5·7, culmin. 0·6, alæ 2·5, caudæ 2·25, tarsi 0·8.

Five specimens were procured, and after comparing them with the series of *Vireonidæ* in the British Museum, there is no doubt that the Fernando Noronha bird comes nearest to *V. magister,* of which species the Museum has now a large series from the islands of the Bay of Honduras, presented by Messrs. Salvin and

Bill of *V. gracilirostris.*   Bill of *V. magister.*

Godman. The yellow face and the slender bill distinguish it at a glance from *V. magister.*

Fam. TYRANNIDÆ.

2. ELAINEA RIDLEYANA.

Elainea Ridleyana, *Sharpe,* P. Z. S. 1888, p. 107.

This species has been fully described by me (*l. c.*). Dr. Sclater (Cat. B. Brit. Mus. xiv. p. 139) does not consider it to be very different from *E. pagana,* but the size of the bill is very marked in the insular birds.

[This bird occurred only on the main island and Rat Island as far as we saw, and was very common in the gardens and in the woods. We saw only a few nests, and of these only one was finished and contained an egg, which was destroyed in an attempt to reach the nest. The egg was white with dark red spots. The nest, which was about three inches across, was made of the tendrils of Cucurbitaceæ and a few fine twigs, but lined thickly (and in fact almost entirely constructed in some cases) with the woolly down of the seeds of *Gonolobus micranthus.* It was placed often in the bare branches of a Burra or *Erythrina* tree, or in a Cashewnut-tree.—*H. N. R.*]

Fam. COLUMBIDÆ.

3. ZENAIDA MACULATA.

Zenaida maculata (*V.*), *Scl. & Salv. Nomencl. Av. Neotr.* p. 132 (1873).
Zenaida aurita, *Gray, List Gallinæ etc. Brit. Mus.* p. 14 (1855).
Zenaida noronha, *Gray, List Columbæ*, p. 47 (1856, descr. nullâ).

The bird from Fernando Noronha is merely a small race of the ordinary *Z. maculata* of the South-American continent, with a slightly shorter wing (5·1–5·4 inches) and tail (2·75–3·2); but as some Brazilian specimens are of the same dimensions, I do not see how the idea of a small insular race can be maintained.

[This little Dove is exceedingly common on all the islands where it can find food, and flies about from one island to the other, singly or in flocks of from 2 or 3 to 30. It is very tame, and even when fired at, or alarmed, usually goes but a short distance before settling. The nest is loose in texture, about 6 inches across, and built of small sticks of the *Spermacoce*, vetches, &c., and lined with roots. It is placed often in the bare branches of a Spondias or Burra, with no attempt at concealment. The eggs are two in number, white, blunt at both ends, and about 1¼ inch long. One bird shot off its nest proved to be a male. The convicts catch these birds both for eating and as pets, keeping them in wicker cages. They are fed on the seeds of Cassias and other Leguminosæ and Cucurbitaceæ, and probably the fig and other succulent fruits.—*H. N. R.*]

Fam. LARIDÆ.

4. ANOUS MELANOGENYS, *Gray*; *Sharpe, Phil. Trans.* vol. 168. p. 467 (1879).

Two adults and a young bird agreed perfectly with specimens obtained on St. Paul's Rock by the 'Challenger' Expedition and determined by Mr. Howard Saunders. The young bird is browner than the adult, and has the head sooty brown with some white on the forehead, eyebrows, and occipital region.

[This Noddy was very common on the island, and is called "Viuva preta." A specimen also flew on board the vessel as we were going to Pernambuco from Europe, about a day's steam from Fernando Noronha. The species nests in small colonies on the rocks in various spots, and also in trees in the Sapate. An egg was obtained from a nest on St. Michael's Mount; it was oval and blunt at both ends, $2\frac{1}{8}$ inches long, and about 1 inch through in the thickest part, chalky-white in colour, marked somewhat sparingly with underlying ash-grey, and overlying sienna. A living young bird from the nest was brought to me, but soon died.—*H. N. R.*]

5. GYGIS CANDIDA (*Gm.*); *Sharpe, t. c.* p. 465.

One adult and two young birds. The latter are white like the old birds, but have much smaller bills.

[This is a common bird in many parts of the island, nesting in trees, especially those of the Sapate, where there is a colony near that of the *Anous.* The bird is called " Viuva bianca."—*H. N. R.*]

Fam. PELECANIDÆ.

6. PHAETHON ÆTHEREUS (*L.*); *Scl. & Salv. Nomencl. Av.* p. 124.

Of this Tropic-bird two specimens were procured. It is common on the island, nesting on the Peak and on other rocks and cliffs. An egg was obtained on St. Michael's Mount. The birds were taken in snares by the convicts.

7. SULA LEUCOGASTRA.

An adult and a young bird. This species of Gannet, known as " Mbebu," is a common bird, nesting on cliffs on all the islands. The young are pure white.

Besides these birds we saw several of which no specimens were procured. *Tachypetes aquila* was abundant, nesting on St. Michael's Mount, and a small species of Albatros appeared several times round the island, but kept well out of gun-shot. Three species of Waders were seen :—One, a small Plover, of which we twice saw a flock at San Antonio Bay, and once or twice single birds flying along the coral-reefs ; a bird resembling a Yellowshank, grey and white, of which a pair appeared at San Antonio at the end of our visit ; and a single specimen of a Sandpiper, at the same spot and time. These wading birds were all very shy, in marked contrast to the endemic species, which suggested that they were migrants, and had come from the mainland, where they are more cautious at the sight of man. The last two species appeared on the same day towards the end of our visit, which confirmed the view that they were migrating.

REPTILIA.

By G. A. BOULENGER, F.Z.S.,
Assistant in the Zoological Department, British Museum.

Only three species were found, viz. a Gecko (*Hemidactylus mabouia*, Mor.), a Skink (*Mabuia punctata*, Gray), and an *Amphisbæna*, described below.

The Gecko is of a widely-distributed species, ranging over the greater part of Tropical America and Africa.

The Skink was originally described from two specimens obtained on Fernando Noronha by H.M.S. ' Chanticleer,' but has since been recorded from Demerara. The specimens brought home by Mr. Ridley are 10 in number; two have 36 scales round the body, the others 38; in one specimen the frontonasal touches the rostral and in another the two shields form a narrow suture.

AMPHISBÆNA RIDLEYI, sp. n.

Under this name I propose to designate an *Amphisbæna* of which a specimen, stated to be from Porto Bello, West Indies *, presented by Capt. Austin, R.N., has been in the British Museum for nearly 50 years, and was referred by Gray, Strauch, and myself to *A. cæca*, D. & B. The same species has been found by Mr. Ridley on Fernando Noronha, and on re-examining the question I find that *A. cæca*, which occurs on various West-Indian Islands, but which was unrepresented in the British Museum when I published my Catalogue, must be regarded as distinct from the one with which I have now the pleasure of connecting Mr. Ridley's name.

16 specimens were collected by Mr. Ridley. One has 180 annuli on the body, one 181, one 182, three 183, two 185, two 186, one 187, one 188, one 189, one 190, one 195, and one 196; two have 18 annuli on the tail, eleven 19, and three 20. The "Porto-Bello" specimen has 189 annuli on the belly, and 19 on the tail. The number of annuli in five specimens of *A. cæca* (including the type) recorded by Strauch are respectively 212+15, 215+13, 227+18, 230+16, and 247+15. Duméril and Bibron give 226-329+18. A specimen from Porto Rico, which I owe to the

---

* I am unable to find such a place either in the West Indies or Northern Brazil, but as the other of the two specimens presented by Capt. Austin as from " Porto Bello " belongs to a North-Brazilian species (*Amphisbæna vermicularis*), I entertain little doubt that both were obtained in Brazil.

kindness of Prof. Lütken, has 228+19. Considering that the
number 247+15 given by Strauch is taken from a specimen in
the Paris Museum, received from the Copenhagen Museum as
from the island of St. Thomas, where only *A. fenestrata* (Cope)
= *antillensis*, R. & L., is known to occur, as Prof. Lütken
kindly informs me, it is clear to me that the specimen with 247
annuli belongs to *A. fenestrata*. The number of annuli would
range, in *A. cæca* from 212 to 229, and in the present species
from 180 to 196. According to Strauch, the length of the labial
border of the first labial shield in *A. cæca* is about one half the
length of that of the second; on Peters's figure of the type
specimen, as well as in the Porto Rico specimen before me, it is
about two thirds ; in *A. Ridleyi* both are equal, or the former is
a little longer. The snout is longer and somewhat more prominent, the tail thicker and more obtuse in *A. cæca* than in
A. Ridleyi. The ventral segments of the two median rows are
broader than long in the former species, the coloration of which
is also different. I have therefore no hesitation in establishing
a new species, which may be characterized as follows :—

Præmaxillary teeth 5 or 7, maxillaries 5-5, mandibulars 8-8.
Snout obtusely pointed, slightly prominent. Tail thinner than
the body, tapering. Rostral small, triangular ; nasals forming a
short suture ; a pair of very large præfrontals, followed by a pair
of much smaller frontals ; eye hardly distinguishable through the
ocular ; a postocular, no subocular ; three large upper labials,
the second and third forming a suture with the ocular ; lower
border of second labial as long as or a little longer than that of
the first, in contact with the second lower labial only ; mental
quadrangular, followed by a large seven-sided chin-shield, which
is much longer than broad ; three lower labials, second very large.
180 to 196 annuli on the body and 18 to 20 on the tail : the
divisions of the annuli longer than broad, nearly equilateral on
the middle of the belly, but nowhere broader than long ; 16 to
18 divisions above, and 20 to 24 below the lateral line. Anal
shields six or eight. Præanal pores four. Uniform brown or
dark purplish brown above, pale brown inferiorly.

		millim.
Length to vent	......	250
Tail	......	24
Diameter of body	......	11

PISCES.

By G. A. BOULENGER, F.Z.S.,
Assistant in the Zoological Department, British Museum.

The following marine species were obtained :—

Apogon imberbis, L., *Hæmulon chrysargyreum*, Günther, *Holocentrum longipinne*, C. & V., *Acantharus chirurgus*, Bl., *Dactylopterus volitans*, L., *Gobius soporator*, C. & V., *Salarias atlanticus*, C. & V., *Salarias romerinus*, C. & V., *Clinus nuchipinnis*, Q. & G., *Clinus delalandii*, C. & V., *Gobiesox cephalus*, Lacép., *Pomacentrus leucostictus*, M. & T., *Glyphidodon saxatilis*, L., *Rhomboidichthys lunatus*, L., *Hemirhamphus unifasciatus*, Ranz., *Clupea humeralis*, C. & V., *Muræna pavonina*, Rich., *Muræna vicina*, Cast., *Muræna catenata*, Bl., and the new species described hereafter.

JULIS NORONHANA, sp. n.

D. $\frac{8}{13}$. A. $\frac{2}{11}$. L. lat. 27. L. tr. $\frac{2\frac{1}{2}}{8}$.

Length of head one third of the total (without caudal), or a little less; depth of the body one fourth. Dorsal spines shorter than the rays. The length of the ventrals is two thirds or three fifths that of the pectoral, which is shorter than the head. Caudalis truncate. Upper half of body and caudal blackish, lower half yellowish white (in spirit); a whitish streak along each side of the back, just above the lateral line; dorsal, anal, pectoral, and ventral fins transparent, immaculate; a black spot between the first and third dorsal rays.

Several young specimens, the largest of which measures 60 millim.

The nearest ally of this species appears to be *J. lucasana*.

MOLLUSCA.

By EDGAR A. SMITH, F.Z.S.,
Assistant in the Zoological Department, British Museum.

The total number of Mollusca now known from Fernando Noronha is 80, of which 72 are marine forms, 7 terrestrial, and 1 freshwater.

Previous to this expedition no land or fluviatile species had been collected, and only 28 marine forms, all obtained by the 'Challenger,' have been recorded from this locality. Ten of these were also collected by Mr. Ridley, who has now added 44 additional species to the list.

The general facies of the marine Molluscan fauna is quite of a West-Indian type, as a perusal of the following pages will show; and it will also be observed that some of the species have a much wider and in some instances a very peculiar range. Of the land-shells two are known West-Indian species, one has been recorded from Brazil, Peru, and the island of Opara, and the remaining four, up to the present, appear to be peculiar to the island. One of these, however, *Bulimus Ramagei*, suggests a faunistic similarity to Brazil, as the section of *Bulimus* to which it belongs (*Tomigerus*), with one exception, occurs only in that country.

The single freshwater species suggests no relationship with any particular region, and might exist anywhere, similar forms being found both in the Old and New Worlds.

The following pages contain an account of the species obtained by Mr. H. N. Ridley and his colleagues, after which is appended a list of those recorded in the 'Challenger' Reports.

I. MARINE SPECIES.

1. OCTOPUS RUGOSUS, *Bosc.*

Hab. Mediterranean, Cape Verde Islands, West Indies, Rio Janeiro, &c.

This species is common in pools at low water. After being dried in the sun the arms are made into soup and eaten by the natives. Mr. Ridley, however, informs me that it is comparatively tasteless and of a soft gelatinous consistency.

2. CONUS NEBULOSUS, *Solander.*

Hab. West Indies : Barbados, Cuba, Martinique, Sta. Lucia.

The operculum of a shell 65 millim. in length is 17 long and only 4½ wide. It is thickened and carinate along the middle beneath, the muscular scar occupying more than half the entire length, and the nucleus is *not terminal* as stated by Messrs. Adams [*] and Tryon [†] in their respective Manuals of Conchology, but situated three millim. from the extremity. The growth at first is regularly concentric, but subsequently, to suit the nar-rowness of the aperture of the shell, the layers of increase are

---

[*] 'Genera of Recent Mollusca,' vol. i. p. 246.

[†] 'Structural and Systematic Conchology,' vol. ii. p. 187.

added at one end only, thus producing a long narrow operculum.
MM. Cross and Marie* have also noticed, in respect of *C. impe-
rialis*, *C. lividus*, and *C. rattus*, that the nucleus of the operculum
is subapical, and doubtless it has a similar position in other
species. The description of the operculum therefore as usually
given in manuals and other works requires modification, and the
nucleus should be termed *apical* or *subapical*.

3. CONUS DAUCUS, *Hwass.*

Hab. Barbados (*Mus. Cuming*); St. Domingo and Guadaloupe
(*Küster*); Cuba and Martinique (*d'Orbigny*).

The single beach-rolled specimen has a very strongly marked
double zone of brown spots upon the middle of the body-whorl.
With this species I unite *C. mammillaris*, Green, *C. castus*, Reeve
(not *C. castus* of Weinkauff), *C. archetypus*, Crosse, and *C. san-
guinolentus* of Reeve.

C. Reevei, Kiener, placed by Weinkauff† in the synonymy of
this species, is quite a distinct shell, which I regard as the same as
C. piperatus, Dillwyn, not *C. piperatus*, Reeve, which, as stated
by Weinkauff, is the same as *C. erythræensis* of Beck.

4. PLEUROTOMA (CRASSISPIRA) FUSCESCENS, *Gray.*

1843. Pleurotoma fuscescens, *Gray, Reeve, Con. Icon.* fig. 125.
1845. Pleurotoma nigrescens, *Gray, Reeve, l. c.* fig. 235.
1845. Pleurotoma paxillus, *Reeve,* fig. 285.
1850. Pleurotoma solida, *C. B. Adams, Contrib. Conch.* vol. i. p. 61.

Hab. Cuba (*d'Orbigny*); Jamaica (*C. B. Ad.* for *solida*); St.
Vincent (*Reeve* for *nigrescens*).

Pl. nigrescens and *Pl. paxillus* differ from the typical form of
the species in being very much smaller, *Pl. solida* being intermediate in size.

In his ' Manual of Conchology ' (vol. vi. p. 193) Tryon states
that *Pl. nigrescens* of C. B. Adams and *Pl. nigrescens* of Gray
are the same species. Having types of the former received from
Adams and Gray's types also for comparison, I can state that
beyond a doubt they are distinct. *Pl. cuprea*, Reeve, is rather
an unsatisfactory species at present, and I am rather inclined to
believe that, as suggested by Tryon, it will prove to belong to
this species also.

* Journ. de Conch. 1874, pp. 333-359.
† Conch.-Cab. p. 312, no. 53.

5. MUREX (OCINEBRA) ALVEATUS, *Kiener.*

Hab. Panama (*Reeve, Kobelt, Sowerby*); West Indies (*Tryon*).

As suggested by Tryon [*], I think there must be some mistake with regard to the locality " Panama " which has been assigned to this species first of all by Reeve and afterwards by others. I have never seen a specimen from that locality, and Mr. G. B. Sowerby informs me that he has frequently received it with collections from the West Indies, but never from the Pacific side of Central America. *M. erosus*, Broderip, *M. obeliscus*, A. Adams, *Triton Cantrainei*, Récluz, and probably *M. pauperculus*, C. B. Adams, are perfectly distinct from the present species and from one another. This is another example of Tryon's rash and indiscriminate "lumping" of species, which detracts so much from whatever value may be attached to his work.

6. PISANIA PUSIO (*Linné*).

Buccinum pusio, Reeve, Con. Icon. fig. 43.

Hab. Honduras and St. Thomas (*Coll. Cuming*); Sta. Lucia (*d'Orbigny* as *Purpura accincta*); Ascension I. (*Conry*).

The specimens from Fernando Noronha are rather small, and much more distinctly striated than certain examples from the West Indies.

7. PURPURA HÆMASTOMA, *Linné.*

The specimens obtained by Mr. Ridley constitute a well-marked variety of this well-known species, both as regards form and colour. They have the spire more elevated in proportion to the length of the aperture, and only the two uppermost of the four series of nodules on the body-whorl are distinct. The interior of the aperture is greenish blue, reddish near the labrum, which is bordered within with black-brown, upon which the fine orange or yellowish liræ are very distinct. The exterior of the shell is purplish black, streaked and spotted with greenish white. *P. hæmastoma* is known from the West Indies, West Africa, Mediterranean, Atlantic coasts of France, Spain, and Portugal.

8. COLUMBELLA MERCATORIA, *Linné.*

Hab. St. Vincent, Grenada, Nevis, Cuba, Martinique, and Sta. Lucia (*Brit. Mus.*).

With one exception the eleven Fernando Noronha shells are

* Man. Conch. vol. ii. p. 128.

white, variegated with very dark brown or black. The single specimen, which differs from the rest, is of a pinkish tint sparingly marked with rich brown.

9. OLIVA LITERATA, *Lamarck*.

Hab. West Indies, Gulf of Florida.

Two beach-rolled shells are all that were obtained. They have the transverse liræ on the inner lip extending over nearly the entire length of the columella.

10. OLIVA (OLIVELLA) NIVEA (*Gmelin*).

Hab. St. Vincent and other islands of the West Indies, Venezuela, and Brazil.

The shell named by Mr. Watson * *Oliva fulgida*, Reeve, from Fernando Noronha, does not belong to that species, but is a prettily coloured example of *O. nivea*. *O. fulgida* differs from Gmelin's species in the form of the columella and basal cauda. The columella of *O. nivea* is very peculiarly excavated, and this may be seen by looking as far within the aperture as possible. No such excavation occurs in *O. fulgida*, which also does not exhibit the numerous oblique folds or liræ on the columellar margin of the aperture which distinguish *O. nivea*.

A second species is quoted with doubt by Watson from Fernando Noronha, namely *O. pulchella*, Duclos. The two fragments referred to this species seem to me to bear little resemblance to Duclos's figure; but 1 have no hesitation in considering them specifically identical with the other specimen from the same spot which I refer to *O. nivea*.

11. LEUCOZONIA CINGULIFERA (*Lamarck*).

Hab. West Indies, Honduras, West Africa.

L. rudis, Reeve, is I consider quite distinct from this species. With this exception I agree with Tryon in his synonymy, and would even suggest the propriety of maintaining *L. leucozonalis*, Lamk., as a variety of this species.

The specimens from Fernando Noronha have stout rounded ribs, exhibit a distinct submedian white zone on the body-whorl, and have the aperture inclining to orange.

Tryon questions the West-African habitat of this species, but I am inclined to think it correct, as in the British Museum there are three specimens from that locality presented some years ago

* Gasteropoda of the ' Challenger' Exped. p. 224.

by a Mr. Lewis, together with other species which are undoubtedly
West-African forms.

12. LEUCOZONIA OCELLATA (*Gmelin*).

Hab. West Indies.

The specimens obtained offer no differences from ordinary
West-Indian examples.

13. LATIRUS SPADICEUS (*Reeve*).

A single young shell seems to belong to this species.

L. concentricus, Reeve, *L. brevicauda*, Reeve, *L. gracilis*, Reeve,
and the present species are very closely related.

14. MITRA BARBADENSIS (*Gmelin*).

Hab. West Indies, Barbados, St. Vincent, &c.

M. tessellata, Kiener, which Reeve named *M. picta*, is perfectly
distinct from the present species, and well known as a South-
African shell. Tryon *, not possessing or not having seen the
species, at once concludes, from their general superficial resemblance, that it must be the same as *M. barbadensis*. The sculpture
of the two is quite different. *M. barbadensis* is ornamented with
raised spiral lines, whilst *M. picta* exhibits transverse punctured
striæ. The character of the outer lip also is quite different.

15. MITRA (PUSIA) ANSULATA, *Sowerby*.

Mitra ansulata, *Sowerby, Thes. Conch.* vol. iv. p. 26, pl. 373. fig. 474.

Mitra microzouias, *Reeve (non Lamarck), Con. Icon.* figs. 185, 202;
Sowerby, l. c. fig. 635; *Kiener, Coq. Viv.* pl. 28. fig. 89 (probably);
Tryon, Man. Conch. iv. p. 183, pl. 54. figs. 568, 569.

Hab. St. Thomas (*Mus. Cuming*); " West Indies, Mörch, Krebs,
and Swift," *fide Tryon.*

This species is said to occur in Polynesia, but the British
Museum Collection affords no evidence in proof of this statement.

The shell from Fernando Noronha belongs to that form of the
species as figured by Reeve (fig. 185).

This is usually considered the *M. microzonias* of Lamarck, but
if it be compared with the figure of that species in the 'Ency-
clopédie Méthodique' (pl. 374. fig. 8), it will be seen that it is
a much more slender shell. The true *M. microzonias* has also
been figured by Küster (Con.-Cab. pl. 17. figs. 12, 13), and Reeve
also correctly depicts it (Con. Icon. pl. xxx. fig. 242 on left) under
the name of *M. leucodesma*. Sowerby in his description of

* Man. Conch. vol. iv. p. 118.

M. ansulata does not mention the presence of a second white zone on the body-whorl as represented in his figure, but this does occasionally exist. Tryon places this species in the synonymy of *M. dermestina*, Lamk., together with *M. cavea*, Reeve, *M. Adamsi*, Dohrn, *M. pulchella*, Reeve, *M. pisolina*, Lamk., *M. histrio*, Reeve, and *M. consanguinea*, Reeve. A more *ridiculous* instance than this of the "lumping" of species I have never seen. Tryon never could have examined examples of these various forms, for if he had he would not have united them; he must have been misled by the figures, or perhaps a little jealousy of non-possession may have influenced him.

Reeve's *M. leucodesma* he says is beyond a doubt the same as *M. pardalis*, Küster. From this it is, in my judgment, perfectly distinct; and the statement that "Reeve's figure of *M. pardalis* is a *Columbella*" is sheer guess-work. The shell figured by Reeve is in the British Museum, and not only is it a *Mitra*, but correctly identified by Reeve as *M. pardalis*, Küster. What right had Tryon to make such a statement in the face of Reeve's description, in which he properly characterizes the shell as *a Mitra with four plaits* on the columella? Numbers of similar absurdities occur throughout this work of Tryon's, which might have been avoided if more judgment had been used and the love of "lumping" been overcome.

16. MARGINELLA SAGITTATA, *Hinds*.

Hab. Bahamas to Brazil.

M. fluctuata, C. B. Adams, from Jamaica, appears to be the same as this species.

17. TRITON RIDLEYI, sp. n.   (Plate XXX. fig. 1.)

Testa late fusiformis, albida, obsolete trizonata, zonis supra varices aurantiis; anfractus normales, superne declives et leviter concavi, ad medium biangulati, inferne constricti, costis longitu-dinalibus circiter 7 (in anfract. ultimo subtuberculiformibus inferne evanescentibus) instructi, liris spiralibus tenuibus (in anfr. pen-ultimo 7–8) aliisque longitudinalibus tenuioribus concinne can-cellati; apertura ovalis, alba; canalis brevis, dextrorsum versus; columella alba, superne arcuata, vix tortuosa, tuberculis vel liris transversis supra callum tenuem munita; labrum compresse varicosum, intus liris duodecim in paribus ordinatis instructum.

Longit. 19 millim., diam. 10.

This species belongs to the same group as *T. gallinago*, Reeve (Con. Icon. fig. 5), and *T. testudinarius*, Adams & Reeve, and some others. Although possibly not adult, the single shell at hand is in excellent condition, and affords all the necessary characters distinctive of the species. The last whorl has two varices, namely the labrum and one on the opposite side. The nuclear whorls are broken off; but, judging from the top of the first normal whorl, the apex would be comparatively small. The uppermost of the liræ on the columella is rather conspicuous, and, together with the uppermost of those within the labrum, forms a semicircular sinus above.

18. TRITON PILEARIS, *Lamarck*.

Hab. West Indies, Red Sea, Ceylon, Philippine Islands, island of Anna, &c. (*Brit. Mus.*).

This, like some other species of *Triton*, occurs at the West Indies and in the Indian and Pacific Oceans.

19. TRITON (EPIDROMUS) TESTACEUS, *Mörch*.

Hab. West Indies (*Mörch*).

This species is very like *T. obscurus*, Reeve, but differs in having more convex whorls, a granulated columellar callus, and a narrower labral varix which is also hollowed out behind.

20. CYPRÆA CINEREA, *Gmelin*, var.

Hab. West Indies.

With this species I unite *C. clara* of Gaskoin, with which the specimens from Fernando Noronha agree. This variety is of a longer and more cylindrical form than the type, has only traces of the black dotting around the base, and no purplish stain between the teeth. Sowerby's figure (Thes. Conch. pl. 307. f. 91*), badly copied by Tryon (Man. Conch. vol. vii. pl. 1. f. 8), does not represent the variety *clara*; but a fair representation of it is given by Sowerby on pl. 316, figure 222. The colour, however, is not pinkish, and no dotting occurs along the sides in the types described by Gaskoin.

21. CYPRÆA (TRIVIA) PEDICULUS, *Linné*.

Hab. West Indies.

One of the specimens from Fernando Noronha is remarkably small, measuring only 7 millim. in length.

22. LITTORINA TROCHIFORMIS, var. ? (Plate XXX. fig. 2.)

Littorina trochiformis, *Dillwyn, Philippi, Abbild.* vol. ii. p. 143, pl. ii. ff. 12, 14, 15.

Littorina nodulosa, *Watson (non Gmelin),* 'Challenger' *Gasteropoda,* p. 577.

Testa parva, fusiformi-ovata, grisea vel nigrescens, albo-nodosa; anfractus 6–7, convexiusculi, superiores granorum seriebus tribus ornati, striisque spiralibus elevatis paucis sculpti, ultimus in medio obtuse angulatus, seriebus quatuor cinctus, ad basim albo punctatus; apertura nigra, fascia basali pallida ornata, inferne subacuminata; columella lata, purpurea, superne macula lutescenti notata.

Longit. 19 mill., diam. 10. Apertura 7 longa, 6 lata.
    ,,   11 ,,   ,, 9.   ,, $6\frac{1}{2}$,, 5 ,,

The above measurements of two specimens from Fernando Noronha show the variation in the form of this variety. The white tubercles are rather acute in some specimens, whilst in others they are scarcely raised above the surface. On the body-whorl there are two *approximated* series at the periphery and two above, and at the base is a tesselation of white and dark spots.

The shells quoted by Mr. Watson from Fernando Noronha are certainly specifically the same as those obtained by Mr. Ridley, and are, I think, almost specifically distinct from the *L. nodulosa* of d'Orbigny. They have less angular whorls and less acute nodules, of which there are *two* series on the body-whorl above the two principal series at the periphery, whilst in *L. trochiformis* (=*nodulosa*, d'Orb.) there is only a single series. The aperture, also, of the Fernando shells is darker and none of them exhibit a second pale zone at the upper part, which is nearly always visible in the West-Indian species.

23. LITTORINA ANGULIFERA (*Lamarck*).

Litorina angulifera, *Philippi, Abbild.* vol. ii. p. 223, pl. v. ff. 12–15.

Hab. West Indies, West Africa and Pacific (*Phil.*).

Only a single young specimen was obtained by Mr. Ridley; it agrees in all particulars with West-Indian specimens.

24. TORINIA ÆTHIOPS (*Menke*).

Hab. West Indies.

Both Philippi and Hanley, in their respective monographs, admit this as a distinct species, but I am inclined to think with the

38*

former * that *T. cyclostoma*, *T. nubila*, *T. cylindracea*, and the present species are mere varieties of one and the same form.

25. IANTHINA FRAGILIS, *Lamarck*.

Of the various species figured by Reeve, that which he has identified as Lamarck's *I. fragilis* (Conch. Icon. pl. ii. ff. 6 *a*, 6 *b*) closely resembles the shells from Fernando Noronha. They have the same perpendicular columella and the same division of colour, the " deep-violet " tint of the base terminating abruptly at the periphery.

26. CERITHIUM ATRATUM (*Born*).

Hab. West Indies, Pernambuco, and Rio Janeiro (*Brit. Mus.*).

I regard the *C. caudatum* of Sowerby as undoubtedly belonging to this species. " Sicily," the locality assigned by Sowerby in the ' Thesaurus Conchyliorum ' and in Reeve's ' Conchologia Iconica ' to *C. atratum*, is evidently incorrect.

27. MITRULARIA ALVEOLATA (*A. Adams*).

Calyptræa alveolata, *A. Adams, Reeve, Con. Icon.* vol. xi. pl. 3. ff. 8 *a-b*.

Hab. Galapagos Islands (*Reeve*); St. Kitts, West Indies (*Brit. Mus.*).

The single shell from Fernando Noronha possesses all the characteristics of the type from the Galapagos Islands. The specimens from Fernando Noronha, assigned with doubt to *M. uncinata*, Reeve, by Watson †, in all probability belong to the same species as that collected by Mr. Ridley. They are, however, only young specimens, so that their determination is all the more difficult.

28. HIPPONYX ANTIQUATUS (*Linné*).

Hipponyx antiquatus (*L.*), *Fischer, Journ. de Conch.* vol. x. p. 5, pl. ii. ff. 1-9 (anatomy); *Crosse, Journ. de Conch.* vol. x. p. 20; *Mörch, Malak. Blätt.* vol. xxiv. p. 98.

Hab. West Indies ; islands of Ascension, St. Helena, and Trinidad in the South Atlantic, Peru and California (*Brit. Mus.*).

The single specimen from Fernando Noronha has the spire more recurved than any other specimen I have seen and it is inclined to the left.

---

* Conch.-Cab., *Solarium*, p. 26.

† ' Challenger ' Gastropoda, p. 461.

29. HIPPONYX GRAYANUS, var.

Hipponyx Grayanus, *Menke, Carpenter, Proc. Zool. Soc.* 1856, p. 4; *Crosse, Journ. de Conch.* 1862, vol. x. p. 23.

The distribution of this species appears to be very extensive. Carpenter quotes it from Galapagos, Sandwich Islands, Panama, S.W. Mexico, Mazatlan, and St. Vincent (W. Africa). Some specimens in the British Museum from St. Helena, wrongly named *H. radiatus*, Quoy & Gaimard, by Jeffreys [*], and two specimens from Fernando Noronha agree in all respects with this species except in the more excentric position of the apex, which gives them a more capuliform appearance.

30. NERITA ASCENSIONIS, *Gmelin*.

In his monograph of *Nerita* in the Conchylien-Cabinet, Dr. von Martens mentions only the island of Ascension and Guinea as localities for this species. I had previously noted [†] the fact of its occurrence at the island of Trinidad off the Brazilian coast, and now I record its presence at Fernando Noronha, where it was also obtained by the 'Challenger' Expedition.

31. TURBO (CALCAR) OLFERSI, *Troschel*.

Trochus Olfersi, *Troschel, Philippi, Conch.-Cab.* ed. 2, p. 126, pl. 22. f. 1.

Calcar Olfersi, *Fischer in Kiener's Coq. Viv.* p. 18, pl. 77. f. 1.

Trochus digitatus, *Reeve (non Deshayes), Conch. Icon.* pl. 5. f. 24; *Sowerby, Thes.* vol. v. pl. 504. fig. 135.

Hab. Brazil (*Philippi & Fischer*).

The localities quoted by Reeve and Sowerby, namely Central America and Panama, will doubtless prove incorrect.

Failing to recognize this species as *T. Olfersi*, Sowerby has placed that name among the synonymy of *T. imbricatus*, which, however, is a perfectly distinct shell. *T. digitatus* of Deshayes, as pointed out by Philippi, Fischer, and Carpenter, is identical with the common *T. unguis*, Wood, of the Californian coast.

32. TROCHUS (EUTROCHUS) JUJUBINUS, *Gmelin*.

Hab. West Indies (*Philippi & Fischer*).

The two specimens from Fernando Noronha are more widely

* Ann. & Mag. Nat. Hist. 1872, vol. ix. p. 264.

† *Ibid.* 1881, vol. viii. p. 431.

umbilicated than the shells figured by Reeve * and Fischer †, and also differ in form, being wider at the base and more shortly conical.

Fischer has already pointed out that the localities of Reeve and Lamarck, Swan River and Mauritius, are probably incorrect.

33. TROCHUS (EUTROCHUS) GEMMOSUS, *Reeve.*

This I believe, as in the case of the preceding species, is another instance of a wrong locality (Philippine Islands) assigned by Reeve.

Two specimens from Fernando Noronha agree in every minute detail with the types in the British Museum, and, as the sculpture and lineations are so remarkable, the identity is beyond doubt. The umbilicus, which is as large as that of *T. jujubinus*, at once distinguishes this species from *T. nobilis*, with which Philippi ‡ questioned its relationship. The type of *Eutrochus* was named *E. perspectivus* by A. Adams ; but as that name was previously used by Koch for another species belonging to the same group, Pilsbury has renamed it *E. Adamsi.*

34. FISSURELLA CANCELLATA, *Sowerby.*

Fissurella cancellata, *Sowerby, Conch. Ill.* sp. 38, pl. 72. f. 29.

Hab. West Indies, Honduras.

With this species I would unite *Fiss. suffusa*, Reeve, and *F. lentiginosa*, Reeve. A third species of the same author, *F. ægis*, is also very similar, but the form is a little more elongate and the radiating riblets are squamose at the points of intersection with the concentric liræ.

35. FISSURELLA ALTERNATA, *Say.*

Fissurella alternata, *Say, Journ. Acad. N. Sci. Philad.* 1822, vol. ii. p. 224 ; *Reeve, Con. Icon.* pl. xii. f. 84 (probably).

Fissurella larva, *Reeve, l. c.* f. 98.

Fissurella Dysoni, *Reeve, l. c.* f. 86.

Hab. Maryland, &c. (*Say*); Bermuda, St. Johns, Honduras (*Brit. Mus.*).

The sculpture of the three above-named forms is essentially he same and the character of the orifice is similar, and all have

---

* Con. Icon., *Zizyphinus*, pl. 2. fig. 12.
† Kiener's Coq. Viv., *Trochus*, pl. 18. f. 2.
‡ Conch.-Cab., *Trochus*, p. 86.

the interior at the apex indented with a transverse line or pit at the larger end of the perforation, as described by Say.

36. ? FISSURELLA BARBADENSIS, *Gmelin*.
Hab. West Indies.

There are two or three specimens from Fernando Noronha which closely approach this species, but I do not feel absolutely certain of the identification.

37. ? FISSURELLA NUBECULA, *Linné*.
Hab. Mediterranean, Spain, Morocco, coast of Gambia, Cape Verd Islands.

Several specimens from Fernando Noronha in some respects so closely resemble this species that I hesitate to separate them. The interior is of the same greenish tint, the orifice has a purplish tint or is ringed with purple, but the outer surface is uniformly darker than Mediterranean examples. With regard to sculpture it is difficult to say that any material difference exists, as specimens from any given locality exhibit slight variations in the fineness and number of the radiating striæ, such as may be noted in the series from Fernando Noronha.

38. ACMÆA NORONHENSIS, sp. n. (Plate XXX. figs. 3, 3 *a*.)

Testa ovata, postice latior, mediocriter elevata, nigrescens, radiis pallidis picta, ad apicem, paulo ante medium situm, erosa, nigra, radiatim tenuiter striata, lineisque incrementi sculpta; pagina interna intra cicatricem nigricans, apicem versus callo tenui sensim albicans, extra cicatricem fere ad marginem cæruleo-albida, ad marginem anguste nigro limbata, antice ab apice usque ad marginem radio lato obscuro et postice alio latiore picta.

Long. 24 millim., lat. 19, alt. 9.

This species has a smoother surface than *A. subrugosa*, d'Orbigny (= *Lottia onychina*, Gould), from Rio Janeiro. Like that species, however, it has in the interior a broad obscure ray from the apex to the margin in front and a broader one at the opposite end. These rays, however, are more distinct in the present species than in the Brazilian shell. The external radiating striæ being very fine, do not, as a rule, produce a crenulated margin, but in some instances a slight crenulation occurs. The surface within the muscular scar is almost black, forming a marked contrast to the pallid space between it and the black margin. The shells found attached to rocks, when placed upon a flat surface, rest upon the anterior and posterior margins only, so that the sides are slightly raised.

39. CHITON (ISCHNOCHITON?) PECTINATUS, *Sowerby*.

Chiton pectinatus, Sowerby, Con. Ill. pl. 174. f. 1-16; Reeve, Con. Icon. pl. 26. f. 133.

Hab. ——? (*Reeve*); West Indies (*P. P. Carpenter in Brit. Mus.*).

The marginal scales are not at all well drawn by Sowerby, being much too elongate.

40. CHITON (ISCHNOCHITON) CARIBBÆORUM, *Carpenter.* (Plate XXX. figs. 5, 5*a*.)

Testa elongato-ovalis, vix carinata, varie picta, grisco-olivacea, albo, rufo et olivaceo picta, vel purpurea, interdum nigrescens, albo virgata, valvis terminalibus concentrice et rugose granoso-striatis vel squamatis, centralibus liris tenuissimis granosis curvatis flexuosis ornatis, areis lateralibus rugose granosis vel squamatis; valva postica pone apicem centralem leviter concava; cingulum minute squamatum, squamis minutis elongatis ovalibus indutum, pallide roseo-griseum, dilute nigro tessellatum.

Longit. 27 millim., diam. 9.

Hab. St. Thomas (*Brit. Mus.*).

The above appears to be a manuscript name attached to specimens in Cuming's collection which are identical with few shells from Fernando Noronha. The colour is very variable, some specimens, when viewed from a distance of twelve inches, appearing olive-grey speckled with white; others are of a pinkish cream-colour speckled with red or blotched along the sides in front of the lateral areas with black, as in some of the specimens from Fernando Noronha. The granules or scales of the lateral area and on the front and posterior valves are peculiarly flat and are somewhat transversely arranged on the former and concentrically on the latter. The central areas are finely punctured along the centre, and become more and more coarsely granosely lirate as the sides are approached. One example is almost entirely reddish purple, and others are blackish with a broad pallid stripe down the middle from end to end.

41. CHITON (ACANTHOCHITON) ASTRIGER, *Reeve*.

Chiton astriger, Reeve, Conch. Icon. pl. xviii. f. 109.

Hab. Barbados.

Reeve describes this species as " smooth along the summit, very closely finely striated on each side." This is not at all accurate. The central portion of the non-terminal valves has a defined elongate subtriangular space which is sculptured with minutely granular lines, and the sides are densely but rather more coarsely granulated. The figure (47) of the detail of sculpture of *C. spiculosa*, Reeve, which I believe to be the same species, gives quite as good an idea of the ornamentation as figure 109. The outer margin of the mantle bears a fringe of the same glassy spicules as compose the tufts.

42. DORIS, sp.

A single specimen was obtained, which appears to belong to the same species as an unnamed example in the British Museum from the West Indies.

43. APLYSIA, sp.

An animal about an inch long is all that was found. It probably is not full-grown. No attempt has been made to identify either this or the preceding, as both belong to difficult groups requiring special study.

44. SIPHONARIA PICTA, var.?* (Plate XXX. figs. 4–4 b.)
Hab. Rio Janeiro (*d'Orbigny*).

The specimens from Fernando Noronha are externally blackish with numerous white radiating costæ. The inner surface also is much darker than in the type specimens from Brazil. *S. hispida*, Gould, also from Rio Janeiro, appears to be the same species. *S. lineolata*, d'Orbigny, from Cuba, is a larger form, and the only species that appears to have been discovered in the West Indies. One or two species are known from Bermuda (*S. brunnea*, Hanley); and Say has described a form from East Florida.

45. VENUS (ANAITIS) PAPILIA, *Linné*.
Hab. West Indies and Cape Verd Islands (*Brit. Mus.*).

46. CARDIUM SUBELONGATUM, *Sowerby*.
Hab. St. Thomas, West Indies.

47. CARDIUM MEDIUM, *Linné*.
Hab. West Indies.

* D'Orbigny, 'Voy. Amér. Mérid.' vol. v. p. 469, pl. 56. ff. 7–11.

48. SEMELE CORDIFORMIS, *Chemnitz.*

1766. Tellina reticulata, *Linné?, Syst. Nat.* ed. 12, p. 1119.

1795. Tellina cordiformis, *Chemnitz, Conch.-Cab.* vol. xi. p. 208, pl. 199. ff. 1941-2.

1815. Tellina decussata, *Wood, Gen. Conch.* p. 190, pl. 43. figs. 2, 3.

1822. Amphidesma orbiculata, *Say, Journ. Acad. Nat. Sci. Philad.* vol. ii. p. 307; *Reeve,* f. 13.

1826. Amphidesma radiata, *Say, l. c.* vol. v. p. 220; *Reeve* (as of *Rüppell*), f. 12.

1832. Amphidesma lenticularis, *Sowerby, Proc. Zool. Soc.* 1832, p. 200; *Con. Ill.* f. 9; *Reeve,* f. 39.

1841. Amphidesma reticulata, *Sowerby, Con. Illust.* p. 8; *Reeve,* f. 29.

1841. Amphidesma subtruncata, *Sowerby, l. c.* p. 7.

1845. Amphidesma Jayanum, *C. B. Adams, Proc. Bost. Soc. Nat. Hist* vol. ii. p. 10.

1853. Semele sinensis, *A. Adams, Proc. Zool. Soc.* 1853, p. 95; *Reeve,* f. 28.

1853. Semele luteola, *A. Adams, l. c.* p. 95; *Reeve,* f. 42.

1853. Semele modesta, *A. Adams, l. c.* p. 95; *Reeve,* ff. 35 *a-b.*

1853. Amphidesma cordiformis, *Reeve, Con. Icon.* f. 30.

Hab. West Indies, Bermuda, Rio Janeiro, Ascension Island, St. Helena, Cape Palmas, and Fernando Po (*Brit. Mus.*).

Having carefully studied a large series of specimens from the above localities, the types of *S. lenticularis,* said to have been obtained in West Colombia, also the types of *S. chinensis, S. reticulata* (Sow.), *S. subtruncata, S. luteola,* and *S. modesta,* also the specimens figured by Reeve, I am inclined to believe that all the above-named forms constitute but a single variable species.

Some authors have considered this species to be the *Tellina reticulata* of Linné; but I am rather inclined to think, with Hanley, that there is not sufficient evidence to determine this with any degree of certainty. The locality " China," from which Adams named a form *S. sinensis,* I regard simply as one more of the innumerable errors of " habitat " occurring in Cuming's collection.

49. CHAMA, sp. incert.

Several specimens of a species of *Chama* were taken on the rocks; but the surfaces are so eroded and water-worn, that it is impossible to determine them.

50. MYTILUS EXUSTUS (*Lamarck*), *Reeve.*

This is a West-Indian species, and was also obtained by the 'Challenger' expedition at Fernando Noronha and Pernambuco *.

51. ARCA IMBRICATA, *Bruguière.*

This species was also obtained by the 'Challenger' Expedition at Fernando Noronha; and other examples were dredged near Cape York, N. Australia. This seems a remarkable distribution; still, as far as I can discover, there appears to be no difference in the shells.

52. ARCA (ACAR) ADAMSII, *Shuttleworth, MS.?.*—Arca Adamsi *in Cuming's Collection.* (Plate XXX. figs. 6, 6 *a.*)

Testa oblonga, subquadrata, sordide albida, inæquilateralis, antice curvata, postice oblique arcuata, inferne in medio levissime sinuata, lineis elevatis radiantibus aliisque concentricis cancellata; umbones parvi, parum remoti, paulo ante medium collocati; area dorsalis augusta, utrinque acuminata; ligamentum minimum, adamantiforme, transversim striatum; pagina interna alba, radiatim plus minus substriata.

Longit. 12 millim., alt. 7½, diam. 7½.

Hab. St. Vincents, Jamaica, and St. Thomas (*Brit. Mus.*).

Two specimens bearing the above name occur in Cuming's collection, and others from Jamaica presented by Dr. P. P. Carpenter are also similarly labelled; but I have not succeeded in finding any description by Shuttleworth of this species.

It is closely related to *A. lactea*, Linn., *A. solida*, Sowerby, and some others. The points of contact of the radiating and concentric liræ are nodulous, and a little coarser than in either of the above-named species. The muscular impressions are clearly defined by a raised ridge which is continued upward towards the umbones.

53. LIMA SQUAMOSA, *Lamarck*, var.

Hab. Atlantic, Pacific, and Indian Oceans.

The distribution of this species is given in my Report upon the 'Challenger' Pelecypoda. The specimens from Fernando Noronha seem intermediate between *L. squamosa* and *L. multicostata*, having fewer ribs than the latter, and more than typical examples of the former.

A specimen 22 millim. long has twenty-six ribs, and another example 16 millim. in length has but twenty-one.

* *Vide* 'Report on 'Challenger' Lamellibranchiata, p. 272.

54. SPONDYLUS, sp.

A few odd valves picked up on the shores are too much worn
to be identified with certainty. About a dozen forms have been
described as West-Indian; and doubtless it is one or more of
these species which occur at Fernando Noronha.

II. TERRESTRIAL SPECIES.

1. HELIX(OPHIOGYRA?) QUINQUELIRATA. (Pl. XXX. figs. 7–7c.)

Testa discoidea, supra leviter convexa, inferne anguste umbilicata, tenuis, viridi-flavescens, nitida; anfractus 7, lente accres-
centes, convexi, incrementi lineis striati, supra peripheriam in-
distincte concavus, antice haud descendens, intus plicis tribus
inæqualibus perlucentibus munitus, lamellisque duobus validis
parietalibus instructus; apertura semilunata, parva; peristoma
tenue, haud expansum.

Diam. maj. 6 millim., min. 5½, alt. 3.

Hab. Found, both living and dead, at the north end of the
island, also on Platform Island.

H. entodonta, Pfeiffer, from Ecuador, is an allied form; but
has a flatter spine, more open umbilicus, and no parietal liræ.

2. BULIMUS (TOMIGERUS) RAMAGEI, sp. n. (Plate XXX. fig. 8.)

Testa subovata, rimata, solida, fusca, zonis angustis albis transversis (in anfr. ultimo quatuor) cincta; anfractus 5, convexiusculi,
lineis incrementi subrugosis, striisque tenuissimis spiralibus
sculpti, ultimus magnus, antice descendens, post labrum con-
tractus, scrobiculatus; apertura irregularis, longitudinis totius ½
paulo superans, dentibus quatuor inæqualibus (duobus parvis in
pariete aperturali, uno magno compresso in margine dextro, uno
tuberculiformi valido in margine columellari) munita; peristoma
album, valde incrassatum, leviter reflexum, marginibus callo craso
junctis.

Longit. 23½ millim., diam. 16.

    „   17½   „   „   12½.

The above measurements show that considerable difference
exists in the size of specimens; and it is a curious fact that the
smallest example, obtained from a native, is the only one which
appears in fairly fresh condition. All the rest were found by
Mr. Ramage imbedded in sandy mud on a raised reef at Tobacco
Point, and have a semi-fossilized appearance.

The only forms at all approaching that now described are the species of *Tomigerus*, all of which, however, have a much more complicated oral dentition, and are of different form, with the last whorl ascending, and other marks of distinctness.

In general form and texture of the shell it is not at all unlike *Pythia inflata*, Pfeiffer; but of course has not the expanded lip or the same dentition as that genus.

Two only of the twenty specimens which I have examined exhibit any variation in the teeth of the aperture. These want the two parietal denticles.

3. BULIMUS (BULIMULUS) RIDLEYI, sp. n. (Plate XXX. fig. 9.)

Testa parva, ovata, superne acuminata, umbilicata, fusca, ad peripheriam luteo-lineata: anfractus 5–6, convexiusculi, lineis incrementi striisque spiralibus tenuissimis sculpti, ultimus antice haud descendens; apertura ovata, intus fusco-carnea, linea pallide mediana ornata, longit. totius $\frac{1}{2}$ ad æquans; peristoma undique expansum, carneo-albidum, marginibus callo tenui (interdum crasso) superne subtuberculiformi junctis.

Longit. 12 millim., diam. 6.

Hab. Living under bark of Mango-trees in the garden and on the north side of the island; also found at the base of the Peak, north side, under stones, and on Rat Island.

I do not know any species sufficiently near this form wherewith to offer a comparison. It resembles somewhat in form certain species of *Partula*; it faintly recalls, chiefly on account of colour, *Bulimus Jacobi*, from the Galapagos Islands; and the spiral striation, although finer, somewhat resembles that of some of the species of the genus *Plecotrema*.

4. PUPA SOLITARIA, sp. n. (Plate XXX. figs. 10, 10 a.)

Testa minuta, rimata, albida, ovato-cylindracea; anfractus 5, convexi, striis incrementi tenuibus sculpti, sutura vix obliqua sejuncti, ultimus pone labrum subvaricosus; apertura parva, quinque-dentata; dente valido duplici columellari, uno minore etiam duplici in pariete aperturali, duobus parvis ab margine dextro remotis, quinto minuto basali; peristoma anguste expansum, album, marginibus callo tenui junctis.

Longit. $2\frac{1}{3}$ millim., diam. $1\frac{1}{4}$.

Hab. Platform Island.

This species is a trifle less cylindrical than *P. pellucida*, Pfeiffer, a Cuban species; but has the number of teeth and their arrangement similar. The columellar tooth, however, is double, the upper portion of it being most prominent; the tooth above it upon the body-whorl is single, bifurcating at the end. The three remaining teeth are rather remote from the margin of the aperture. The anterior part of the last whorl just behind and parallel to the labrum exhibits a longitudinal swelling or varix, towards which the lip expands.

5. STENOGYRA (OPEAS) OCTONOIDES, *C. B. Adams.*
Hab. Jamaica, Cuba, St. Thomas.

This species is more strongly striated than *S. subula*, has rounder whorls, a deeper suture, and a larger apex.

Four specimens were obtained at Platform Island.

6. STENOGYRA (OPEAS) SUBULA, *Pfeiffer.*
Hab. Cuba, Porto Rico, &c.

The specimens from Fernando Noronha agree exactly with examples of this species from Porto Rico. They were found beneath stones on the promontory between Chaloupe Bay and S. Antonio Bay.

7. STENOGYRA (OPEAS) BECKIANA, var.
Hab. Island of Opara, Peru, Brazil.

The shells from Fernando Noronha, obtained at the same spot as the preceding species, answer well to Pfeiffer's description; but they are *more strongly costulate* than Brazilian specimens with which I have compared them.

III. FRESHWATER SPECIES.

1. PLANORBIS NORONHENSIS, sp. n. (Plate XXX. figs. 11–11 *b*.)

Testa parva, valde compressa, superne in medio depressa, inferne subplanulata, albida, tenui, subpellucida; anfractus 4, superne convexiusculi, striis incrementi tenuissimis sculpti, inferne radiatim subplicati, ad suturam angustissime marginati, ultimus infra medium obtuse carinatus; peristoma tenue, marginibus callo filiformi junctis, superiore oblique arcuato.

Diam. maj. 5 millim., min. 4$\frac{1}{3}$, alt. 1.

This species is about the same size and shape as *P. Gilberti*, Dunker, and *P. fragilis*, Brazier, from Australia. The lower surface, however, is flatter, and the curve of the lip different

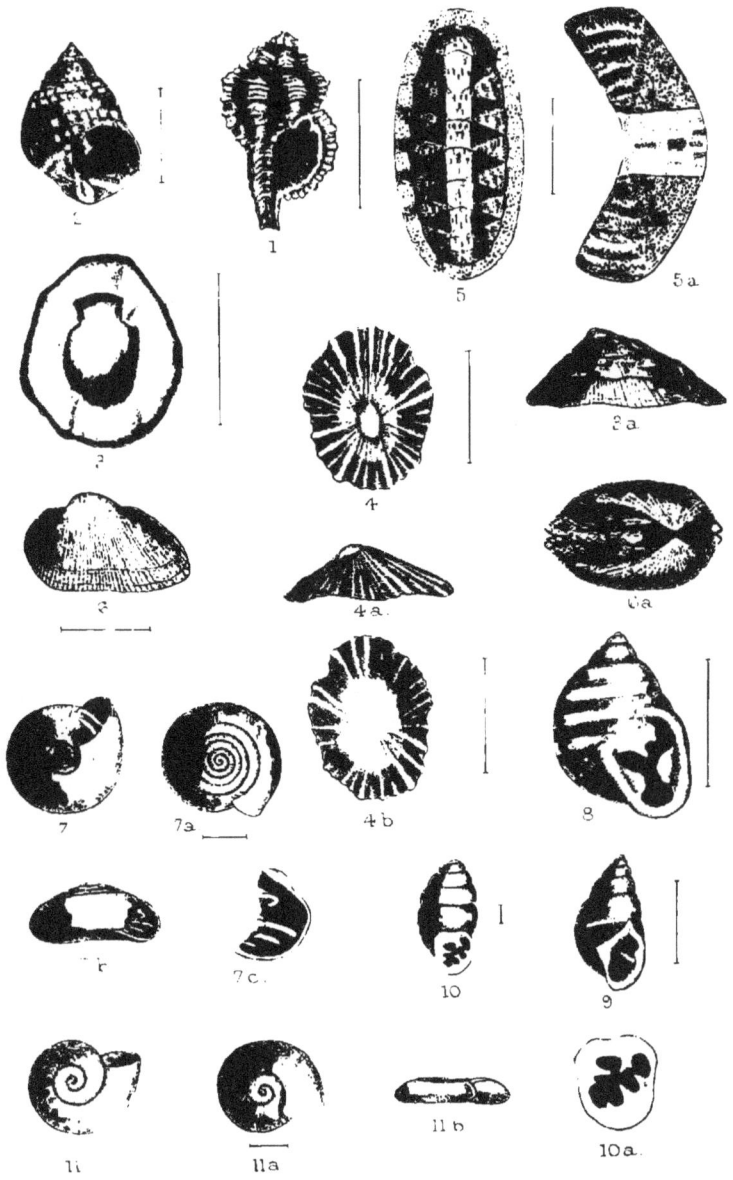

MOLLUSCA OF FERNANDO NORONHA

when viewed from above. It was very plentiful in the lake on the south-west corner of the island.

List of Species obtained at Fernando Noronha by the 'Challenger' Expedition.

Acmæa, sp.
Littorina nodulosa, *d'Orb.*
Nerita ascensionis, *Gmelin.*
Cerithiopsis, sp.
Columbella mercatoria, *Linn.*
Cylichna noronyensis, *Watson.*
Fossarus ambiguus (*Linn.*).
Marginella sagittata, *Hinds.*
Mitrularia uncinata (*Rve.*).
Nassa capillaris, *Watson.*
Oliva fulgida, *Reeve.*
—— pulchella (?), *Duclos.*
Phasianella, sp.
Rissoa, sp.

Scalaria hellenica, *Forbes.*
Siphonodentalium tetraschistum, *Watson.*
Solarium, sp.
Stomatella nigra, *Quoy & G.*
Utriculus canaliculatus (*Say*).
Xenophora corrugata (*Reeve*).
Chiton Boogii, *Haddon.*
Pectunculus pectinatus (*Gmelin*).
Ervilia subcancellata, *Smith.*
Cardium medium, *Linné.*
Lucina pecten, *Lamarck.*
Mytilus exustus, *Rve.*
Arca imbricata, *Brug.*
Pecten noronhensis, *Smith.*

For the above species, see the Reports on the Gasteropoda, Polyplacophora and Lamellibranchiata, by R. B. Watson, A. C. Haddon, and E. A. Smith respectively.

EXPLANATION OF PLATE XXX.

Fig.   1. *Triton Ridleyi,* sp. n.
    2. *Littorina trochiformis,* var.
  3, 3 a. *Acmæa noronhensis,* sp. n.
  4–4 b. *Siphonaria picta,* var.
    5. *Chiton (Ischnochiton) carribæorum.*
    5 a. Ditto. Central valve, magnified.
  6, 6 a. *Arca (Acar) Adamsii.*
  7–7 c. *Helix (Ophiogyra?) quinquelirata.* 7 c. Aperture, enlarged; lip broken away to show the teeth.
    8. *Bulimus (Tomigerus) Ramagei,* sp. n.
    9.   ,,   (*Bulimulus*) *Ridleyi,* sp. n.
   10. *Pupa solitaria,* sp. n. 10 a. Aperture, enlarged.
11–11 b. *Planorbis noronhensis,* sp. n.

POLYZOA.

By R. KIRKPATRICK,
Assistant in Zoological Department, British Museum.

POLYZOA.

The specimens chiefly encrust shells, and are generally much worn away.

1. AETEA RECTA, *Hincks*.

2. SYNNOTUM AVICULARE, *Pieper*.

3. SCRUPOCELLARIA FRONDIS, n. sp.

4. CRIBRILINA RADIATA, *Moll*.

5. SMITTIPORA ANTIQUA, *Busk*.
   (*Mollia antiqua*, Smitt.)

6. STEGANOPORELLA SMITTII, *Hincks*.

7. MASTIGOPHORA DUTERTREI, *Audouin*.

8. SCHIZOPORELLA UNICORNIS, *Johnston*.

9. LEPRALIA DEPRESSA, *Busk*.
   (*Escharella rostrigera*, Smitt.)

10. LEPRALIA CLEIDOSTOMA, *Smitt*.

11. RHYNCHOPORA BISPINOSA, *Johnst.*—(Encrusting *Gorgonia* axis.)

12. CELLEPORA RIDLEYI, n. sp.

13. MICROPORELLA VIOLACEA, *Johnst.*—Encrusting *Gorgonia* axis; (both purple and white varieties).

14. CRISIA HOLDSWORTHII, *Busk*.

15. AMATHIA BRASILIENSIS, *Busk*.

Family CELLULARIIDÆ.

Genus SCRUPOCELLARIA.

SCRUPOCELLARIA FRONDIS, n. sp.

Zoœcia of medium size, alternate; area oval, occupying nearly half the front of cell; spines 2–4 on the outer side, 2 on the inner; lowermost spine on outer side bending over the top of the aperture and giving off processes from its upper border; operculum entire, large, oval, marked with concentric striæ; on

Fig. 1.

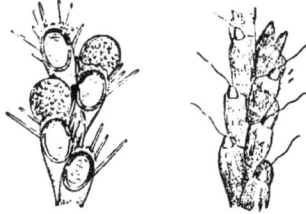

Scrupocellaria frondis, n. sp.

some cells a small pointed avicularian cell projecting from the front of the cell; lateral avicularia wanting; on dorsal surface vibracula, small, flattened, obliquely placed; setæ long. Oœcia small, globose, vitreous, punctured.

The presence of the antler-like spine across the top of the area is a marked character of *S. frondis*.

Loc. Fernando Noronha; Pernambuco.

Family CELLEPORIDÆ.

Genus CELLEPORA.

CELLEPORA RIDLEYI, n. sp. (Fig. 2, p. 506.)

Zoarium loosely encrusting; zoœcia decumbent, rectangular, and flattened at the margins, heaped, somewhat ventricose sub-vertical in the centre, separated by raised lines; orifice from semicircular to subquadrate, with concave proximal margin; two or three short processes surrounding the orifice; at base of an anterior process a small avicularium facing inwards, with small semicircular mandible. On the front of some cells a small avicularian cell with small rounded mandible. Oœcium shaped like a thick semi-disc, concave below, overhanging the mouth of the

cell ; mucronate processes on the upper surface; front wall
of oœcium with a semicircular membranous area on the front
wall.

Fig. 2.

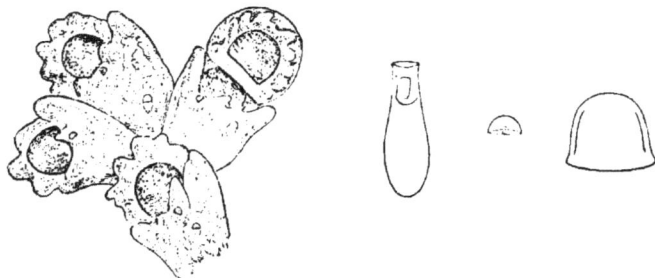

Cellepora Ridleyi, n. sp.

The oœcium of *C. Ridleyi* is remarkable in its shape and relations
to the zoœcium ; also there is a curious resemblance between the
membranous area and the orifice of the zoœcium.

Loc. Fernando Noronha.

CRUSTACEA.

By R. I. POCOCK,
Assistant in the Zoological Department, British Museum.

Introductory Remarks.—The fauna is in all essential respects
allied to that of the mainland and of the Antilles.   The following
wide-spread forms were, as might have been expected, met with:—
*Grapsus maculatus, Leiolophus planissimus, Hippa scutellata,
Alpheus Edwardsii,* and *Gonodactylus chiragra.* There are two
new species of *Alpheus,* one of *Panulirus,* and one of *Stenopusculus*
(*S. spinosus*).   The last mentioned genus has hitherto only been
known from the island of Mauritius ; its occurrence here, therefore, is of great interest.   A new freshwater Ostracod was also
obtained.

DECAPODA.

MAIOIDEA.

Family PERICERIDÆ.

Genus MICROPHRYS, *M.-Edwards*.

1851. Microphrys, *M.-Edwards, Ann. Sci. Nat. Zool.* 3, xvi. p. 251.
1879. Microphrys, *Miers, Journ. Linn. Soc.* (*Zool.*) xiv. p. 664.
1881. Microphrys, *A. M.-Edwards, Miss. Sci. Mex.* (*Crust.*) p. 59.

MICROPHRYS BICORNUTUS (*Latreille*).

1825. Pisa bicornuta, *Latreille, Encycl. Méth. Hist. Nat.* x. p. 141.
1872. Microphrys bicornutus, *A. M.-Edwards, Nouv. Arch. Mus. Hist. Nat.* viii. p. 247.
1881. Microphrys bicornutus, *id. Miss. Sci. Mex.* (*Crust.*) p. 61, pl. xiv. figs. 2, 3, 4.

Nine specimens, six males and three females (two with ova).

This species is common on the coasts of Florida, Mexico, and of the West-Indian Islands. Occurred under stones and on coral-reef.

Genus MITHRAX (*Leach*).

1817. Mithrax (*Leach*), *Latreille, Règne Animal*, iii. p. 23.
1834. Mithrax, *Milne-Edwards* (in pt.), *Hist. Nat. Crust.* i. p. 317.
1879. Mithrax, *Miers, Journ. Linn. Soc.* (*Zool.*) xiv. p. 667.

MITHRAX VERRUCOSUS, *M.-Edwards*.

1832-38. Mithrax verrucosus, *M.-Edwards, Mag. Zool.* vii. pl. 4.
1881. Mithrax verrucosus, *M.-Edwards, Miss. Sci. Mex.* p. 102.

Four specimens, two males and two females (one with ova).

The largest specimen (a male), with the following measurements of carapace, width 42 mm., length 35 mm., differs considerably from the others, of which the smallest (the female with ova) gives the following measurements of carapace :—width 17 mm., length 15 mm. In the three small specimens all the spines are sharper and relatively longer, and the carpus of the chelipedes is armed above with four or five minute spines in addition to the three spines which adorn its anterior (interior) margin.

Brazil and the West Indies are localities given for this species. Under stones at Morro do Chapeo.

MITHRAX (TELEOPHRYS) CRISTULIPES (*Stimpson*).

1862. Teleophrys cristulipes, *Stimpson, Ann. Lyc. Nat. Hist.* vii. p. 190 pl. ii. fig. 2.

1881. Teleophrys cristulipes, *A. M.-Edwards, Miss. Sci. Mex.* (*Crustacea*), p. 113, pl. xix. fig. 2.

Regions of the carapace defined by shallow sulci. Superior surface of body and limbs tubercular, inferior surface smooth.

Carapace broader than long, beset with low, inconspicuous, scattered tubercles which vary in distinctness. The principal tubercles arranged as follows:—two or three on each half of the rostrum between the superior orbital prominences, one on each side of the middle line at the base of the rostrum, two in longitudinal series on each side of the gastric region, several on the branchial regions, and three on the anterior lateral margin of each. Orbit furnished in front with a superior and an inferior blunt prominence. Not furnished with spines or prominences behind.

Rostrum short, broad, with upturned anterior margin, not deeply bifid, marked above with central sulcus, and separated on each side from the superior orbital prominence by a conspicuous depression. Its anterior margin projecting slightly beyond the middle of the basal segment of the antennæ and slightly in front of the inferior orbital prominence.

Chelipedes large ; merus tubercular above and furnished below in front with three large, rounded, compressed teeth ; carpus furnished above with four or five tubercles, and with one blunt tooth in front. Hand smooth; its distal portion compressed above and below into a crest. Dactylus and pollex meeting only at the apices. Dactylus furnished with a single tooth.

In the first pair of legs the merus is furnished above with two longitudinal rows of prominences, the posterior row consisting of lower rounded tubercles, which distally decrease in size, the anterior row of five higher, compressed, sharper teeth, which distally increase in size ; distal margin of the segment produced into five rounded prominences, varying in size ; the carpus, in addition to three or four low tubercles on its centre, with its distal margin furnished with a larger anterior and a smaller posterior tubercle, and its antero-superior surface with a larger proximal and a smaller distal tooth; propodos furnished above with two tubercles, one near the centre, the other at its distal margin ; claw long, curved, hairy below, with its distal portion serrate below.

The arrangement of tubercles and teeth upon the second, third, and fourth pairs of legs is nearly the same as the arrangement upon the first pair, but the posterior row of meral tubercles becomes progressively fainter from before backwards, and the teeth of the anterior row become gradually modified in form and number until, in the posterior pair of limbs, this row is formed of four teeth, two larger and two smaller, the larger and smaller alternating, and one of the larger being the most proximal of the series. Width of carapace $7\frac{1}{2}$ mm., length 7 mm.

One male specimen was obtained.

To guide me in the identification of the Fernando-Noronha specimen, which I refer to *T. cristulipes* (Stimps.), I have had to trust to the descriptions and figures of that species published by Dr. Stimpson and by M. Alphonse Milne-Edwards, and to my own examination of a single imperfect individual which was taken off Cape St. Lucas (California), and presented to the British Museum by the Smithsonian Institute.

Now, although with the above-mentioned figures and descriptions the specimen from Fernando Noronha does not present agreement in all points, yet, making allowance for possible errors on the part of the artists, I should unhesitatingly have referred this specimen to *T. cristulipes* (Stmps.) were it not for the fact that the points of difference between it and the specimen from Cape St. Lucas are by no means inconsiderable.

In the Californian specimen the sulci defining the regions of the carapace are conspicuously deeper, and the tubercles of the same part, though exhibiting in the main the same arrangement, are much larger. This is especially the case with regard to those of the branchial region, the three low tubercles of the antero-lateral margin in the Noronha specimen being represented in the Californian specimen by three large upstanding teeth. Again, with regard to the limbs, the merus of the chelipede in the Californian specimen is furnished below in front with one large compressed tooth and the pollex is armed with two small teeth, these small teeth being scarcely represented in the Noronha specimen. The other limbs present much the same arrangement of teeth in the two specimens, but, as in the case of the carapace, the teeth of the Californian specimen are relatively larger than those of the Noronha specimen.

I am well aware that the differences thus set forth are amply sufficient to justify the separation as distinct species of the specimens which they characterize; yet having but one example from each locality, I am unable to determine the constancy of the differences presented, and must consequently leave the decision of the question as to the specific identity or distinction of the two to those whom either the possession of a long series of forms or a more perfect acquaintance with this group of Crustacea places in a better position to judge than myself.

I am not aware that this species, or at all events any closely allied form, has before this been recorded from the eastern coast of America. Stimpson obtained it from Cape St. Lucas and M. Alphonse Milne-Edwards has described it from the Bay of Panama.

MITHRAX (MITHRACULUS) CORONATUS (*Herbst*).

1782. Cancer coronatus, *Herbst, Naturg. der Krabben*, i. p. 184, pl. xi. fig. 63.

1881. Mithraculus coronatus, *A. Milne-Edwards, Miss. Sci. Mex.* (*Crustacea*), p. 106, pl. xx. fig. 1.

Eight males and seven females (three with ova) were taken.

This species occurs on the coasts of Brazil, Central America, and of the West-Indian Islands. Its presence in Fernando Norouha has been previously mentioned by Mr. E. J. Miers, two small specimens having been obtained from that island during the voyage of H.M.S. 'Challenger.'

CANCROIDEA.

Family CANCRIDÆ.

Genus CARPILIUS (*Leach, MS.*), *Desmarest*.

1825. Carpilius, *Desmarest, Consid. gén. sur la classe des Crust.*, footnote, p. 104.

1834. Carpilius, *Milne-Edwards* (pt.), *Hist. Nat. Crust.* i. p. 380.

1865. Carpilius, *A. Milne-Edwards* (pt.), *Nouv. Arch. Mus. Hist. Nat.* i. p. 212.

1886. Carpilius, *E. J. Miers, Brachyura of H.M.S. 'Challenger,'* p. 110.

CARPILIUS CORALLINUS (*Herbst*).

1782. Cancer corallinus, *Herbst, Naturg. der Krabben*, i. p. 133, pl. v. fig. 40.

1865. Carpilius corallinus, *A. Milne-Edwards, Nouv. Arch. Mus. Hist. Nat.* i. p. 216.

Three specimens, two males and one female.

This species is the West-Indian representative of the genus. [These crabs are exported in wooden crates filled with dry leaves to Pernambuco, where they are in great demand as food. We were told they were land-crabs.—*H. N. R.*]

Genus ACTÆA, *de Haan.*

1850. Actæa, *de Haan, Crust. in Siebold, Fauna Japonica,* dec. i. p. 18.

ACTÆA ACANTHA, *Milne-Edwards.*

1834. Cancer acanthus, *M.-Edwards, Hist. Nat. Crust.* i. p. 379.

1881. Actæa acantha, *A. M.-Edwards, Miss. Sci. Mex. (Crust.)* p. 245, pl. xliii. fig. 1.

One minute specimen, a male, was obtained.

It is only comparatively lately that the locality of this species has been made known by M. Alphonse Milne-Edwards, who received a specimen of it from Guadeloupe.

Genus LEPTODIUS, *A. Milne-Edwards.*

1863. Leptodius, *A. Milne-Edwards, Ann. Sci. Nat., Zool.* sér. 4, xx. p. 283.

1873. Leptodius, *A. Milne-Edwards, Nouv. Arch. Mus. Hist. Nat.* ix. p. 221.

1886. Leptodius, *Miers, Brachyura of H.M.S. ' Challenger,'* p. 136.

LEPTODIUS AMERICANUS (*Saussure*).

1858. Chlorodius americanus, *H. de Saussure, Mém. sur divers Crust. nouv. du Mex. et des Antilles,* p. 14, pl. i. fig. 5.

1881. Leptodius americanus, *A. Milne-Edwards, Miss. Sci. Mex. (Crust.)* p. 269.

Of this species eight specimens (5 males, 3 females with ova) were obtained. It is found in the West Indies and Florida.

Genus LOPHACTÆA, *A. Milne-Edwards.*

1862. Lophactæa, *A. Milne-Edwards, Ann. Sci. Nat., Zool.* sér. 4, xviii. p. 43.

1865. Lophactæa, *A. Milne-Edwards, Nouv. Arch. Mus. Hist. Nat.* i. p. 245.

1886. Lophactæa, *E. J. Miers, Brachyura of H.M.S. ' Challenger,'* p. 113.

LOPHACTÆA LOBATA, *Milne-Edwards.*

1834. Cancer lobatus, *Milne-Edwards, Hist. Nat. Crust.* i. p. 375.

1865. Lophactæa lobata, *A. Milne-Edwards, Nouv. Arch. Mus. Hist. Nat.* i. p. 249, pl. xvi. fig. 3.

A single male specimen. This is a West-Indian and Mexican species.

GRAPSOIDEA.

Family OCYPODIDÆ.

Genus OCYPODA, *Fabricius.*

1798. Ocypoda, *Fabricius* (pt.), *Ent. Syst. Suppl.* p. 317.
1837. Ocypoda, *Milne-Edwards, Hist. Nat. Crust.* vol. ii. p. 41.
1880. Ocypoda, *Kingsley, Proc. Acad. Nat. Sci. Philad.* p. 179.
1886. Ocypoda, *Miers, Brachyura of H.M.S. 'Challenger,'* p. 237.

OCYPODA ARENARIA (*Catesby*).

1771. Cancer arenarius, *Catesby, Hist. of the Carolinas,* ii. p. 35, pl. xxxv.
1880. Ocypoda arenarius, *Kingsley, Proc. Acad. Nat. Sci. Philad.* p. 184.
1882. Ocypoda arenarius, *Miers, Ann. Mag. Nat. Hist.* ser. 5, x. p. 384, pl. xvii. fig. 7.

Five adult specimens (four males and one female).

[Common in holes in the sand at Sueste Bay and Sambaquichaba.—*H. N. R.*]

Family GRAPSIDÆ.

Genus GRAPSUS, *Lamarck.*

1818. Grapsus, *Lamarck* (pt.), *Hist. Nat. Anim. sans Vert.* v. p. 247.
1880. Grapsus, *Kingsley, Proc. Acad. Nat. Sci. Philad.* p. 192.
1886. Grapsus, *Miers, Brachyura of H.M.S. 'Challenger,'* p. 254.

GRAPSUS MACULATUS (*Catesby*).

1771. Pagurus maculatus, *Catesby, Nat. Hist. Carolinas,* ii. p. 36, pl. xxxvi. fig. 1.
1880. Grapsus maculatus, *Kingsley, Proc. Acad. Nat. Sci. Philad.* p. 192.

This species has a very wide range, occurring upon the coasts of the warmer temperate and tropical parts of the Indian, Pacific, and Atlantic Oceans. It is exceedingly variable, and the extent of variation is well shown by the specimens brought from Fernando Noronha.

Five immature and two adult males were obtained; the former

are of a dark green colour with feeble indications of *maculæ*, the latter red-brown with maculæ well developed. [Very common on the rocks all over the group, running briskly just above water-mark and leaping from stone to stone.—*H. N. R.*]

Genus PLAGUSIA, *Latreille*.

1806. Plagusia, *Latreille* (pt.), *Gen. Crust. Ins.* i. p. 33.
1837. Plagusia, *Milne-Edwards, Hist. Nat. Crust.* ii. p. 90.
1878. Plagusia, *Miers, Ann. Mag. Nat. Hist.* ser. 5, i. p. 148.
1886. Plagusia, *Miers, Brachyura of H.M.S. ' Challenger,'* p. 271.

PLAGUSIA DEPRESSA (*Fabricius*).

1775. Cancer depressus, *Fabricius, Syst. Ent.* p. 406.
1782. Cancer squamosus, *Herbst, Naturg. der Krabben,* i. p. 260, pl. xx. fig. 113.
1878. Plagusia depressa, *Miers, Ann. Mag. Nat. Hist.* ser. 5, i. p. 149.

[This ran about on the stones and rocks like the *Grapsus.—H. N. R.*]

Genus LEIOLOPHUS, *Miers*.

1850. Acanthopus, *de Haan, Faun. Japon., Crust.,* p. 29 (nom. præocc.).
1876. Leiolophus, *Miers, Cat. New-Zeal. Crust.* p. 46.
1878. Leiolophus, *Miers, Ann. Mag. Nat. Hist.* ser. 5, i. p. 153.

LEIOLOPHUS PLANISSIMUS (*Herbst*).

1804. Cancer planissimus, *Herbst, Naturg. der Krabben,* iii. Heft 4, p. 3, pl. lix. fig. 3.
1878. Leiolophus planissimus, *Miers, Ann. Mag. Nat. Hist.* ser. 5, i. p. 153.

A single specimen (female with ova) of this wide-spread form was taken.

PORCELLANIDEA.

Family PORCELLANIDÆ.

Genus PETROLISTHES, *Stimpson*.

1859. Petrolisthes, *Stimpson, Proc. Acad. Nat. Sci. Philad.* x. p. 227.

PETROLISTHES MARGINATUS, *Stimpson*.

1862. Petrolisthes marginatus, *Stimpson, Ann. Lyc. Nat. Hist. New York,* vii. p. 74.

I have had no opportunity of examining specimens of *P. marginatus* (Stimpson), and consequently not being certain of the correctness of the identification of the specimens that I have referred to that species, I have thought it desirable to publish a description of them which may, so far as is possible, furnish a test as to the accuracy of the conclusion that has been arrived at.

Carapace and upper surface of limbs pubescent. Width of carapace approximately equal to its length. Carapace smooth, punctured; its anterior half furnished laterally with a small, sharp, upstanding spine. From this spine there extends backwards into the posterior half of the carapace a granular ridge which serves to separate the superior portion of the carapace from the lateral portion. The frons is slightly depressed and is marked off from the hinder portion of the carapace by a distinct ridge, which runs transversely between the posterior margins of the orbits. In the middle this ridge is interrupted by a conspicuous sulcus, which extends to the central lobe of the frons. This lobe is rounded anteriorly; its lateral margins are nearly vertical to the remainder of the anterior margin of the frons and approximately parallel to the superior margin of each orbit, which is the lateral border of the frons. The anterior half of this lateral border marked off from the posterior half by being at a conspicuously lower level.

Basal segment of antenna furnished on the inner side with a small acute spine.

Upper surface of chelipede covered with more or less squamiform granules; lower surface smooth. Anterior margin of upper surface of meral segment produced into a sharp process; beneath this, on the under surface, is a sharp spine, which may be bifid; posterior margin of upper surface spined. Anterior margin of upper surface of carpal segment furnished with three or four sharpened processes; posterior margin spined and produced distally into a spined process. The middle of the upper surface bearing a longitudinal series of larger squamiform tubercles. Inferior border of anterior surface of carpal segment granular; rest of the surface smooth. Anterior and posterior margins of manus and dactylus granular. A slightly curved series of larger squamiform granules extending along the upper surface of the manus from its carpal to the middle of its dactylar joint. Continuous with this is a series running from the base to the apex of the dactylus. Apex of dactylus and of thumb smooth and curved.

Anterior and posterior margins of meral segments of second,

third, and fourth pairs of legs spined ; posterior margin of second and third pairs produced distally into a small acute spine.

Colour (of specimens preserved in spirit of wine) red or yellow above, with darker spots, reddish pink beneath.

Three specimens. Length and width of carapace in largest specimen 14 mm. ; length of manus and pollex 20 mm.

The specimens that I have here described and identified provisionally as *P. marginatus* (Stmps.) are evidently closely allied to *P. asiaticus* (Leach), the common Indo-Pacific form, and I am doubtful if they should be regarded other than as varieties of that species.

HIPPIDEA.

Family HIPPIDÆ.

Genus REMIPES, *Latreille*.

1806. Remipes, *Latreille, Gen. Crust. Ins.* i. p. 45.
1837. Remipes, *Milne-Edwards, Hist. Nat. Crust.* ii. p. 204.

REMIPES SCUTELLATUS (*Fabricius*).
1793. Hippa scutellata, *Fabricius, Ent. Syst.* ii. p. 474.
1858. Remipes cubensis, *H. de Saussure, Mém. sur Crust. nouv. du Mex. et des Antilles*, p. 36, pl. ii. fig. 19.
1878. Remipes scutellatus, *Miers, Journ. Linn. Soc. (Zool.)* xiv. p. 319.

The species occurs on the tropical coasts of the Atlantic.

Twenty-three specimens, two of which are females with ova, were taken. [Very common on the sandy shores. When a wave broke, these little crustacea were often seen running and burying themselves in the sand as the water retired.—*H. N. R.*]

THALASSINIDEA.

Family GEBIIDÆ.

Genus GEBIA.

1816. Gebia, *Leach, art.* Annulosa, *Edinb. Encycl.* vii. p. 419.
1837. Gebia, *Milne-Edwards, Hist. Nat. Crust.* ii. p. 312.

GEBIA SPINIGERA, *S. I. Smith*.
1869. Gebia spinigera, *Smith, Rep. Peabody Acad. Sci.* p. 92.

Eight specimens, one female with ova, were brought back.

The species was originally described from specimens obtained upon the west coast of Central America.

ASTACIDEA.

Family PALINURIDÆ.

Genus PANULIRUS, *Gray.*

1847. Panulirus, *Gray, Cat. Brit. Mus. (Crust.)* p. 69.
1852. Panulirus, *Dana, Crust. U.S. Expl. Exp.* i. p. 519.

PANULIRUS ECHINATUS, *S. I. Smith.*

1869. Panulirus echinatus, *Smith, Trans. Connecticut Acad.* ii. p. 20.

Five specimens were taken, two adult females, one with ova, and one immature female, one adult male and one immature male.

The specimens described by Smith were from Pernambuco.

PANULIRUS ORNATUS (*Fabr.*).

1798. Palinurus ornatus, *Fabricius, Ent. Syst. Suppl.* p. 400.
1837. Palinurus ornatus, *M.-Edwards, Hist. Nat. Crust.* ii. p. 296.
1867. Palinurus ornatus, *Heller, Reise Freg. Novara, Crust.* p. 99.

In 1872 v. Martens, in his paper "Ueber cubanische Crustaceen," Arch. f. Naturg. xxxviii. p. 128, recorded the occurrence on the eastern coasts of America of a *Palinurus*, which he questionably identified as *P. ornatus* (Oliv. ?), a species which appears to have its head-quarters in the Indo-Pacific Seas. From Fernando Noronha, Mr. Ridley obtained one specimen of a *Panulirus*, which I cannot separate by any important character from *P. ornatus* (Fabr.); and in addition to this specimen there is in the British Museum Collection one other from Panama, which is also, I believe, referable to *P. ornatus* (Fabr.). It will thus be seen that this form occurs in the Indo-Pacific Seas and upon the east and west coasts of America.

It is perhaps of interest to note that the spines upon the carapace and upon the peduncles of the antennæ appear to be somewhat sharper, and relatively longer, in the American individuals than they are in the Eastern individuals that I have had an opportunity of examining.

[Tolerably common, and collected from the rock-pools for food.—*H. N. R.*]

PANULIRUS INERMIS, n. sp.

Carapace somewhat flattened above, with sides nearly vertical. The right and left portions of the upper surface meeting in the middle line at a very obtuse angle. Carapace nearly smooth frontal spines considerably shorter than the eye-stalks, slightly incurved at the apices, armed above at the base with a single spine; one spine situated near the ocular margin of the carapace, one in the anterior third of the supero-lateral margin, and a third beneath the eye-stalk near the outer portion of the basal antennal segment.

Antennal peduncle about two thirds the length of the carapace; basal segment armed externally with a single spine on its anterior margin; second segment armed above with five spines, two forming a longitudinal series externally, three forming an oblique series internally; third segment armed above with ten short spines. Below, the three segments are smooth.

Antennular plate nearly horizontal, with rounded antero-external angles not armed with spines; the peduncle shorter than peduncle of antennæ; segments of peduncle not spined.

Epistoma with a straight unspined anterior margin.

The first and fifth pairs of limbs simple, unspined. (Second, third, and fourth pairs absent.)

Postero-external angles of the sternum prolonged into a sharp, long spine.

Abdominal tergites smooth, punctured, not marked with a transverse sulcus; inferiorly and laterally prolonged into a spine. The posterior margin of the last dorsal plate furnished with two long, sharp spines.

Proximal portion of telson furnished in the middle of its upper surface with two spines and with its posterior margin armed on each side with four spines.

Total length from anterior margin of carapace to posterior margin of telson 27 millim. Length of upper surface of carapace 11 millim.

One specimen.

Judging from its size, the specimen from which the above description has been taken is certainly immature. It, nevertheless, presents the characters of a true *Panulirus*, and differs from all the specimens of that genus that I have examined in the absence of spines from the basal plate of the antennulæ. Dredged in Water Bay. About 10 fathoms depth.

CARIDEA.

Family PALÆMONIDÆ.

Genus ALPHEUS (*Fabricius*).

1798. Alpheus, *Fabricius, Ent. Syst. Suppl.* p. 380.

1878. Alpheus, *Kingsley, Bull. U.S. Geol. Surv.* iv. p. 189.

ALPHEUS EDWARDSII (*Aud.*).

1809. Athanasus Edwardsii, *Audouin, Explic. planches de Savigny, Descript. de l'Égypte, Atlas,* pl. x. fig. 1.

1818. Alpheus heterochelis, *Say, Journ. Acad. Nat. Sci. Philad.* i. p. 243.

1884. Alpheus Edwardsii, *Miers, Rep. Crust. H.M.S. 'Alert,'* p. 284.

Twenty-nine specimens. This species is common in the warmer parts of the Atlantic, Pacific, and Indian Oceans, and in consequence of its wide range and of the variations to which individuals are subject it possesses a long list of synonyms. These synonyms may be found upon reference to the above cited work of Mr. E. J. Miers.

ALPHEUS MINOR, *Say*.

1818. Alpheus minus, *Say, Journ. Acad. Nat. Sci.* i. p. 245.

1837. Alpheus minus, *Milne-Edwards, Hist. Nat. Crust.* ii. p. 356.

1878. Alpheus minus, *Kingsley, Bull. U.S. Geol. Geogr. Surv.* iv. p. 190.

One specimen.

This species occurs upon the east and west coasts of America. Kingsley records it from N. Carolina, Bermudas, Florida on the east, and from Pearl Island Bay (Panama) on the west.

ALPHEUS RIDLEYI, n. sp.

Carapace and abdominal tergites smooth; carapace furnished in front with a short pointed rostrum, which does not nearly reach to the second segment of the antennular peduncle; rostrum separated by depression from the ocular hoods, each of which is furnished with a spine projecting in front as far as the extremity of rostrum.

Antennular spine reaching nearly to the second segment of the peduncle, which is the longest of the three, the third being the shortest.

Antennal scale as long as antennal peduncle, longer than antennular peduncle; basal segment of antenna furnished beneath with a strong spine.

Terminal segment of external maxillipede hairy.

First pair of legs very unequal in size. Dactylus of larger hand closing vertically, with evenly rounded supero-anterior border, without accessory teeth ; its greatest length equal to one half of the length of the superior margin of the manus. Anterior margin of the "thumb" on the inner side nearly vertical, forming an obtuse angle with the inclined superior margin. Superior and inferior margins of thumbs on the outer side nearly parallel : in front united by a distinct anterior border, which below curving forwards forms with the inferior border the tooth of the thumb, which does not project so far forwards as the anterior margin of the dactylus.

Upper margin of the manus with a very faint constriction in its anterior half ; right and left sides smooth, without depressions ; lower margin with a very faint depression in its anterior half ; upper margin marked with sulcus, which in the middle of the hand curving downwards and backwards runs to the carpal joint. Carpus rounded above, not bearing a tooth ; meros three-sided, flattened below, not bearing a tooth above in front. Smaller manus simple, without constrictions or depressions ; dactylus, thumb, and upper margin of manus approximately equal in length ; carpus furnished with a blunt tooth above, equal in size to the carpus of the larger manus ; meros resembling the meros of the larger manus.

In the second pair of legs the first carpal segment is as long as the second and the third together ; third about half the length of the second, equal in length to the fourth, shorter than the fifth, which itself is shorter than the second.

Meros and carpus of third and fourth pairs of legs not spined.

In size and form resembling *A. Edwardsii*, but differing from it in having the larger hand very lightly constricted above and below. Moreover, there is a large black spine on each side of the telson.

Alpheus panamensis, *Kingsley*.

1878. Alpheus panamensis, *Kingsley, Bull. U.S. Geol. Surv.* iv. p. 192.

Carapace smooth, furnished in front with a strong rostrum, which projects considerably beyond the spines of the orbital hoods, almost as far as the second segment of the antennular peduncle ; separated by a depression from the ocular hoods, each of which is furnished with a small sharp spine.

Lower margin of hood continuous below the spine.

Antennular spine reaching slightly beyond the margin of the basal segment of the peduncle. Second segment of peduncle longer than the third, as long as the first.

Antennal scale and peduncle as long as each other, and slightly longer than the antennular peduncle. Basal segment of antenna furnished below with a strong sharp spine.

First pair of legs very unequal in size. Dactylus of larger hand closing vertically, its greatest length being more than half the length of the superior margin of the manus; without accessory teeth.

Anterior margin of the thumb on the inner side nearly vertical, meeting the inclined superior margin at an obtuse angle; less than half the length of the superior margin. Thumb on the outer side without a vertical anterior margin, the superior margin meeting the inferior at an acute angle and forming the tooth.

Manus smooth, without constrictions or depressions, longer than the carapace; superior and inferior margins nearly parallel. Carpus rounded above, not bearing a tooth. Meros three-sided, flattened below; superior margin produced in front into a blunt process.

Smaller manus simple; dactylus and thumb approximately equal in length to each other and to the manus.

Carpus furnished above on the inner side with a small projection. Meros resembling meros of larger limb, except that the front process is smaller.

In second pair of limbs the carpal segments are 1, 2 and 5, 3 and 4.

First segment almost as long as the second, third, and fourth together. Second segment a little shorter than the third and fourth together, these being approximately equal; fifth as long as the second.

Meros and carpus of third and fourth pairs of legs not spined. Dactyli of limbs not bifid.

One specimen. If I am right in referring this species to *Al. panamensis* of Kingsley, with the description of which it agrees well, it is of interest to note that it occurs upon the eastern and western coasts of America. Mr. Kingsley described his specimens from Panama and Acajutla.

ALPHEUS OBESO-MANUS, *Dana*.

1852. Alpheus obesomanus, *Dana, U.S. Expl. Exped., Crustacea*, i. p. 547, pl. xxxiv. fig. 7.

Carapace smooth, furnished in front with a short rostrum,

which does not reach nearly so far as the anterior border of the first segment of the antennula, but a little beyond the ocular hoods, from which it is separated on each side by a deep sulcus. Ocular hoods not spined, but slightly produced in front.

Antennular spine short, not reaching to the front margin of the first segment of the peduncle of the antennula. Second segment of peduncle the longest of the three, the third the shortest.

Antennal scale as long as peduncle of antennula, shorter than peduncle of antenna. Basal segment of antenna without a spine.

Legs of first pair very unequal in size. Dactylus of larger manus closing horizontally, about half as long as the upper margin of the manus. The superior (outer) margin of the thumb furnished with two large teeth, of which the posterior is smaller, more slender, and with a blunt apex, the anterior having a rounded margin.

Dactylus short, rounded. Dactylus and thumb very hairy.

The manus simple, cylindrical, without constrictions or depressions, as long as the carapace and the two proximal segments of the antennular peduncle. Carpus deep from above downwards, rounded above, and not furnished with a tooth. Meros deep from above downwards, three-sided, flattened below; upper margin produced in front into a conspicuous process.

Smaller hand somewhat resembling the larger, except that it is less twisted, less cylindrical, with dactylus and thumb straighter and relatively longer. Carpus less deep, and furnished on the upper inner margin with a distinct nodule. Meros less deep, with upper tooth scarcely conspicuous.

In the second pair of legs the carpals are 2, 5, 4 and 3 and 1. The first, third, and fourth segments approximately equal in length, the fifth a little longer; the second as long as the third, fourth, and fifth together. In the third and fourth pairs of legs the carpus and meros are below furnished in front with a strong spur.

Ten specimens. So far as I know, this species has not been hitherto recorded from the American coasts. Its occurrence has been mentioned in the Samoan Islands (*Kingsley*), Fiji Islands (*Dana, Miers*), and in Mauritius (*Richters*).

ALPHEUS ROSTRATIPES, n. sp.

Carapace smooth, anterior margin crescentically excavated, the sides of the excavation being formed by the ocular hoods, which are anteriorly produced but not furnished with spines, and not separated from the rostrum by a depression. Rostrum springing from the centre of the excavation, pointed, short, projecting slightly in front of the ocular hoods, but not reaching the anterior margin of the first segment of the antennular peduncle. Antennular spine reaching to the middle of the second segment of peduncle. Segments of peduncle short, approximately equal in length; second segment furnished externally with a small spine on its anterior margin. Antennal scale as long as the antennular peduncle, much shorter than the antennal peduncle. Basal segment of antenna furnished laterally with a conspicuous spine.

One of the legs of the first pair absent. The dactylus of the remaining one (the smaller?) closing vertically; long, longer than the manus, curved, pointed blade-like, when closed : crossing the thumb. Thumb almost as long as dactylus, and at the base twice as thick, gradually tapering to a sharp, upturned point, meeting manus at an obtuse angle. Manus elliptical, simple, without constrictions or depressions, furnished close to the dactylar joint on each side with two blunt teeth, those on the outer side being obscurely marked. Carpus furnished on its inner side with a small blunt tooth. Meros three-sided, flattened below, superior margin produced in front into a conspicuous projection. Carpals of the second pair of legs becoming progressively shorter in the following order :—1, 5, 2, 3, 4.

Carpus and meros of third and fourth pairs not furnished below with a spine. Dactylus of fourth pair bifid, of third pair absent.

ALPHEUS, sp.

(Too mutilated for identification.)

Carapace furnished in front with a small pointed rostrum, which projects slightly in front of the ocular hoods, but not nearly to the anterior margin of the basal segment of the peduncle of the antennula ; ocular hoods rounded and not spined. Basal spine of antennula reaching to the anterior margin of the basal peduncular segment. Second segment of peduncle longer than the third, approximately equal to the first. Antennal scale longer than the antennular peduncle, shorter than the peduncle of the antenna. Basal segment of antenna not provided with a spine.

Legs of first pair absent.

Carpals of the second pair differing upon the two sides, on the right side the fifth segment being longer than the second, and on the left side shorter. In each case the first is the longest, and the third and fourth the shortest.

Carpus and meros of third and fourth pairs not produced below into a tooth.

Dactyli of third, fourth, and fifth pairs simple.

ALPHEUS, sp.

(Too mutilated for identification.)

Carapace furnished in front with a short rostrum, which does not project as far as the middle of the first segment of the peduncle of the antennula, and is separated by a depression on each side from the ocular hoods. Each ocular hood furnished with a spine. Antennular spine short, sharp, not reaching to the front margin of the basal segment of the peduncle. Second segment of the peduncle the longest, the first and third approximately equal in length. Antennal scale as long as antennal peduncle, a little longer than antennular peduncle. Basal segment of antenna furnished with a long, sharp spine, which projects as far as the middle of the second segment of the antennular peduncle.

First and second pairs of legs absent. Carpus and meros of third and fourth pairs not produced below in front into a strong process. Dactyli of third, fourth, and fifth pairs bifid.

[The *Alphei* were taken in numbers from the holes in which they hid by breaking up the coral-reef.—*H. N. R.*]

Family PENÆIDÆ.

Genus STENOPUSCULUS, *Richters*.

1880. Stenopusculus, *Richters, Beiträge zur Meeresfauna der Insel Mauritius und der Seychellen, von Möbius, Richters und v. Martens*, p. 167.

STENOPUSCULUS SPINOSUS, n. sp.

? Syn. Stenopusculus crassimanus, *Richters, t. c.* p. 168, pl. xviii. figs. 27-29.

Upper portion of cephalothorax sparsely spined; spines in
40*

front of the cervical suture larger than those behind it. Posterior margin of the cervical suture furnished above with 4 or 5 spines, and laterally with 3 or 4 larger ones. Postero-lateral portions of cephalothorax almost smooth; antero-lateral portions beset with spines arranged more or less in longitudinal series. Anterior marginal excavation adjoining the basal antennal segment armed with four spines.

Cephalothorax furnished in front with a pointed rostrum, which starts upon the anterior half of the cephalic portion of the carapace and reaches almost as far forwards as the front margin of the antennular peduncle. Upon each side the rostrum extends horizontally over the basal portion of the eye. Furnished above with eleven teeth, of which five are larger than the rest, and below near the apex with one tooth.

Proximal portion of antennular peduncle furnished externally with a strong curved spine; upper surface of peduncle with three spines; under surface with four on the inner margin and one on the outer margin.

Basal segment of antennal peduncle furnished above with two spines externally, and with a laminate process internally; second segment covered by the basal segment, furnished below with three spines; third segment with one spine externally and with three internally. External margin of antennal scale with five or six fine teeth, internal margin fringed; antennal scale somewhat triangular, laminate, projecting slightly in advance of the antennal peduncle, which is approximately as long as the antennular peduncle.

Epistome furnished with four strong teeth.

Ischial segment of external maxillipede furnished distally with three spines externally, and with one spine internally; meral segment externally with three strong spines. Internal margin of all the segments clothed with hairs.

Segments of first and second pairs of legs simple, more or less cylindrical, unspined.

Meropodite of third pair of legs cylindrical, spined, with some larger sharp spines near the distal extremity on the inner surface. Carpopodite rounded below, flattened and hollowed above; the hollowed portion with a few small spines, the rest thickly spined; spines on the outer surface larger.

Inner surface of the hand covered with small tubercles; outer

ZOOLOGY OF FERNANDO NORONHA. 525

surface almost smooth, with a few small tubercles near the upper and under margins. Upper margin compressed into a serrated keel; under margin also serrated. Anterior margin of hand nearly at right angles to the axis of the pollex. Pollex upturned at the apex, furnished on its occludent margin with a tooth which closes behind the tooth of the dactylopodite. Upper margin of dactylopodite serrated.

Fourth and fifth pairs of limbs resembling each other in being slender and elongated, in having the propodite furnished below with a series of fine spines and consisting of three segments, and in having the dactylopodite bifid. But whereas the propodite of the fourth pair consists of five segments, the propodite of the fifth pair consists of but three. The number of divisions of these segments, however, appears to vary upon the two sides.

Abdominal tergites smooth above; lateral portions narrowed, somewhat pointed, and with margins more or less spined.

The outer and inner lamellæ of appendages of the sixth abdominal somite with a median longitudinal crest, serrate exterior margin, and fringed inner margin. Outer margin of inner lamella furnished below with a stronger tooth.

Telson with converging lateral margins, rounded posterior margin; each lateral margin furnished with a central tooth; posterior margin furnished with three teeth, one on each side and one in the middle. Upper surface of telson marked with two longitudinal crests, each of which bears three spines arranged longitudinally; the depression between the crests furnished proximally with four spines in two longitudinal series. Base of telson bearing on each side one marginal spine.

Two specimens.

Length from apex of rostrum to posterior margin of telson 13 millim.; total length of upper surface of carapace (including rostrum) 5½ millim.; length of manus and pollex of third pair of feet 8 millim.

This species seems to differ from *St. crassimanus*, Richters, in the possession of a greater number of teeth upon the rostrum and in the absence of a crest upon the abdominal tergites.

The three species which hitherto have, so far as I am aware, composed the genus were taken at Touquets (Mauritius).

STOMATOPODA.

Genus GONODACTYLUS, *Latreille.*

1825. Gonodactylus, *Latreille, Encycl. Méth. Hist. Nat.* x. p. 473.
1837. Gonodactylus, *Milne-Edwards, Hist. Nat. Crust.* ii. p. 528.
1880. Gonodactylus, *Miers, Ann. Mag. Nat. Hist.* v. p. 115.
1886. Gonodactylus, *Brooks, Stomatopoda of H.M.S. 'Challenger,'* p. 55.

GONODACTYLUS CHIRAGRA (*Fabricius*).

1793. Squilla chiragra, *Fabricius, Ent. Syst.* ii. p. 513.
1880. Gonodactylus chiragra, *Miers, Ann. Mag. Nat. Hist.* v. p. 115.

Fourteen specimens of this widely distributed species were brought back. In the coral-reef.

MYRIOPODA.

By R. I. POCOCK,
Assistant in the Zoological Department, British Museum.

The island does not seem to be rich in members of this group, since four species only were obtained in it. Two of these appear to be new to science, one being referable to the genus *Geophilus*, the other to the genus *Spirobolus*. The others are the two widespread tropical species, *Scolopendra morsitans* (Linn.) and *Paradesmus gracilis* (C. Koch).

CHILOPODA.

SCOLOPENDRA MORSITANS (*Linn.*), emend., *Kohlrausch, Arch. f. Naturg.* 1881. p. 104.

Thirteen specimens were taken. Common under dung and stones, at the east end of the main island and base of Peak Garden and elsewhere. The bite is about as bad as a wasp's sting.

GEOPHILUS RIDLEYI, n. sp.

Length 44 millim. Width about 1 millim. Posterior end of the body slightly more slender than the anterior.

Ochraceous, head-plate slightly darker.

Number of pairs of legs 73 (in the female).

Antennæ hirsute, the distal end more so than the proximal. Segments of the proximal half cylindrical, those of the distal half narrowed proximally ; apical segment as long as the two preceding segments.

Head-plate with straight anterior margin, rounded lateral margins, and concave posterior margin : sparsely clothed with hairs, and almost destitute of punctures.

Frontal lamina coalesced with rest of head-plate.

Basal lamina about twice as wide as long, with abruptly converging lateral margins and concave anterior margin. The prebasal lamina visible in the space left between the concave posterior border of the head-plate and the concave anterior border of the basal lamina.

Maxillary sternite wider than long; its anterior margin slightly excavated, but scarcely bidentate.

Maxillary feet largely visible from above, and projecting slightly in front of the head-plate ; the segments on the inner side furnished with hairs but not armed with teeth.

Dorsal plates conspicuously bisulcated.

Ventral pores occupying a circular area in the posterior half of the sternites.

Legs sparsely clothed with longish hairs.

The anal tergite wide, but not covering the pleuræ ; with rounded postero-lateral angles, staight posterior margin, and lateral margins slightly converging behind.

Anal pleuræ smooth, not furnished with punctures.

Anal sternite very wide at the base, gently converging lateral borders, rounded posterior angles and straight posterior margin.

Anal pores conspicuous. Anal legs broken.

A single female specimen, found under a stone in the Sapate.

This species appears to be closely allied to *G. occidentalis*, Meinert (Proc. Amer. Phil. Soc. xxiii. p. 220), from San Francisco ; but differs in the absence of teeth from the segments of the maxillary feet, and in the absence of pores from the anal pleuræ.

Diplopoda.

PARADESMUS GRACILIS (*C. Koch*).

Two female specimens.

For the synonymy and an excellent description of this species, see Dr. Robert Latzel's 'Die Myriopoden der öst.-ungar. Monarchie,' ii. p. 162.

This very wide-spread form occurs in the East and West Indies and Brazil.

It has been introduced, in connection with tropical plants, into Europe; and I have examined many specimens of all ages, which were captured in the conservatory of Mr. Alfred O. Walker at Chester, and in the orchid-houses of Mr. Herbert Druce at St. John's Wood.

[It was very common in the garden.—*H. N. R.*]

SPIROBOLUS (s. s.) NORONHENSIS, n. sp.

Length about 30 millim. Number of somites 37.

Colour deep slate-grey or almost black; anterior half of each somite (the first and last excepted) adorned above and below on each side with a single reddish spot. Legs and labrum reddish.

Distal portion of the head-plate furnished with a faint median longitudinal impression, upon each side of which, near the margin of the labrum, are two setiferous punctures, one near the middle line, the other near the external portion of the labral excavation. Distal segments of antennæ pilose.

First dorsal plate smooth, without striæ; laterally, where the anterior and posterior margins pass into one another, evenly rounded: furnished with a fine sulcus which runs from near the ocular region of the head-plate, close to the antero-inferior margin, and terminates at the postero-inferior margin.

Foramina repugnatoria situated, somewhat dorsally, in the posterior portion of the somites. Posterior portion of each somite smooth above; anterior portion marked with transverse striæ; inferior and lateral portions marked with numerous longitudinal striæ. Somites not furnished with the '*scobina*.'

Posterior somite smooth; produced behind into a blunt rounded process, which extends slightly beyond the margins of the anal valves. Anal valves with margins not compressed. Margin of subanal plate rounded.

The right and left moieties of the male copulatory apparatus held together in front by a triangular plate. Below this plate terminates in a rounded apex, which extends as far as the inferior margins of the halves of the apparatus. Each upper angle of this plate produced laterally and upwards into a relatively slender bar, which curves round the superior portion of the anterior lamina of its side. Each anterior lamina simple, more or less spatulate, with evenly rounded external margin and slightly concave inferior margin. Viewed from the side, seen to be considerably thicker above than below. Posterior lamina irregularly quadrate, with even outer and inner margins which below slightly converge; the inner margin distally produced into a conspicuous rounded, noduliform process, which projects slightly below the level of the concave inferior margin of the anterior lamina, and is consequently visible when the copulatory apparatus in its entirety is viewed from the front. Above and externally, the interspace between the anterior and posterior laminæ is occupied by a small sclerite, with even margins and rounded below, which, dilating above, forms the posterior margin of the superior aperture of the sheath, of which the walls are composed of the four laminæ just described, and which contains the protrusible portion of the copulatory apparatus. This protrusible portion is articulated at its proximal end to a simple rod, which is itself articulated to the upper extremity of the anterior lamina. Protrusible portion curved almost through the arc of a semicircle, and composed of two segments. The distal segment about twice as long as the proximal, membranous and hollow behind, chitinous in front, with its posterior portion armed with a simple small process.

A dozen specimens found under stones in the Banana plantations at the base of the Peak.

In many points this species appears to resemble *Sp. paraensis* (Humb. & Sauss.). But the absence of all knowledge of the form of the copulatory apparatus of that species makes it impossible for me to refer these specimens to it.

INSECTA, excepting Coleoptera.

By W. F. KIRBY, F.L.S., F.E.S.,
Assistant in the Zoological Department, British Museum.

Notwithstanding the comparatively large proportion of new species in the present collection, it would be a mistake to suppose that very many will ultimately prove to be confined to the island of Fernando Noronha. The greater part belong to Orders of insects which are comparatively little collected or studied, and among which large numbers of conspicuous species remain to be described, even from the best explored tropical countries.

The few Lepidoptera in the collection were taken at an unfavourable season of the year, and many are worn specimens. They exhibit more decidedly West-Indian affinities than might have been expected.

I prefix to the paper a full list of all the species obtained, except a few which were worn, immature, or too scantily represented for satisfactory identification. They are arranged systematically under the orders and principal families to which they belong.

List of Species obtained.

Order ORTHOPTERA.

FORFICULIDÆ.

1. Pygidicrana notigera, *Stål.*
2. Labidura riparia, *Pall.*
3. Anisolabis jancirensis, *Dohrn.*
4. —— Antoni. *Dohrn.*

BLATTIDÆ.

5. Phyllodromia poststriga, *Walk.*
6. Ischnoptera lucida, *Walk.*
7. Periplaneta americana, *Linn.*
8. Blatta incommoda, n. sp.
9. Leucophæa surinamensis, *Linn.*
10. Euthyrrapha pacifica, *Coq.*

GRYLLIDÆ.

11. Scapteriscus abbreviatus, *Scudd.*
12. Gryllus assimilis, *Fabr.*
13. —— forticeps, *Sauss.*
14. (Œcanthus (?) pallidocinctus, n. sp.

PHASOPTERIDÆ.

15. Conocephalus vernalis, n. sp.
—— ——, var. n. frater.
16. Œcella (n. g.) furcifera, n. sp.
17. Meroncidius viridinervis, n. sp.

LOCUSTIDÆ.

18. Stenopola dorsalis, *Thunb.*

Order NEUROPTERA.

ODONATA.

LIBELLULIDÆ.

19. Pantala flavescens, *Fabr.*
20. Tramea basalis, *Burm.*

Order HYMENOPTERA.

TEREBRANTIA.

ENTOMOPHAGA.

CHALCIDIDÆ.

21. Blastophaga obscura, n. sp.
22. Ganosoma dispar, n. sp.

EVANIIDÆ.

23. Evania lævigata, *Latr.*

ACULEATA.

HETEROGYNA.

FORMICIDÆ.

24. Camponotus bimaculatus, *Smith.*
25. Pheidole omnivora, n. sp.

FOSSORES.

BEMBICIDÆ.

26. Monedula signata, *Linn.*

POMPILIDÆ.

27. Pompilus nesophilus, n. sp.

LARRIDÆ.

28. Tachytes inconspicuus, n. sp.

DIPLOPTERA.

VESPIDÆ.

29. Polistes Ridleyi, n. sp.

ANTHOPHILA.

ANDRENIDÆ.

30. Halictus lævipyga, n. sp.
31. —— alternipes, n. sp.
32. —— atripyga, n. sp.

Order LEPIDOPTERA.

RHOPALOCERA.

LYCÆNIDÆ.

33. Tarucus Hanno, *Stoll.*

HETEROCERA.

NOCTUÆ.

34. Heliothis armiger, *Hübn.*
35. Anomis (?) dispartita, *Walk.*
36. Anthophila flammicincta, *Walk.*
37. Bolina bivittata, *Walk.*
38. Thermesia gemmatalis, *Hübn.*

GEOMETRES.

39. Nemoria denticularia, *Walk.*
40. Acidalia Fara, n. sp.

PYRALES.

41. Pyralis manihotalis, *Guén.*
42. Samea castellalis, *Guén.*
43. Hymenia perspectalis, *Hübn.*
44. Phakellura hyalinata, *Linn.*
45. Margaronia jairusalis, *Walk.*
46. Acharana phæopteralis, *Guén.*
47. Pachyzancla detritalis, *Guén.*
48. Opsibotys flavidissimalis, *Grote.*

CRAMBI.

PHYCIDIDÆ.

49. Mella zinckenella, *Treitschke.*

Order **HEMIPTERA.**

HETEROPTERA.

PENTATOMIDÆ.

50. Pentatoma testacea, *Dall.*

LYGÆIDÆ.

51. Lygæus rufoculis, n. sp.

52. Heræus variegatus, n. sp.
53. Ligyrocoris balteatus, *Stål.*
54. —— bipunctatus, n. sp.

VELIIDÆ.

55. Rhagovelia incerta, n. sp.

Order **DIPTERA.**

DOLICHOPODIDÆ.

56. Psilopus metallifer, *Walk.*

SYRPHIDÆ.

57. Temnocera vesiculosa, *Fabr.*

MUSCIDÆ.

58. Sarcophaga calida, *Wied.*

Description of New Species and Special Notes.

1. PYGIDICRANA NOTIGERA, *Stål.*
Pygidicrana notigera, *Stål, Eugenie's Resa, Zool. Ins.* p. 299 (1858).
Flew into light.

2. LABIDURA RIPARIA (*Pall.*).
Forficula riparia, *Pall. Reise,* ii. Anhang, p. 30 (1773).
A cosmopolitan species.

3. ANISOLABIS JANEIRENSIS (*Dohrn*).
Forcinella janeirensis, *Dohrn, Stett. ent. Zeit.* xxv. p. 285.

4. ANISOLABIS ANTONI (*Dohrn*).
Forcinella Antoni, *Dohrn, Stett. ent. Zeit.* xxv. p. 289 (1864).
These earwigs were common under stones in the main island.

5. PHYLLODROMIA POSTSTRIGA (*Walk.*).
Blatta poststriga, *Walk. Cat. Blatt.* p. 99, n. 69 (1868).
The locality of the typical specimen is unknown.

6. ISCHNOPTERA LUCIDA, *Walk.*
Ischnoptera lucida, *Walk. Cat. Blatt.* p. 120, n. 39 (1868).
A single immature specimen, probably belonging to this species.
Taken under stones, base of Peak.

7. PERIPLANETA AMERICANA (*Linn.*).
Blatta americana, *Linn. Syst. Nat.* i. p. 424, n. 4 (1758).
A cosmopolitan species.   Common and introduced.

8. BLATTA INCOMMODA, n. sp.

Long. corp. 11½ millim.

Female. Ferruginous brown; the thorax and sides of the abdomen varied with black; legs and costal margin of the tegmina testaceous. Pronotum rather long, moderately convex, the sides converging in front, the hinder angles rounded off, and the hind border convex. Tegmina broad, covering the whole base of the abdomen, but ceasing at about two fifths of its length.

Similar to *B. orientalis,* Linn., but much smaller, the tegmina much larger, and the pronotum longer.

9. LEUCOPHÆA SURINAMENSIS (*Linn.*).

Blatta surinamensis, *Linn. Syst. Nat.* i. p. 424, n. 3 (1758).

A cosmopolitan species. Under stones, base of Peak.

10. EUTHYRRAPHA PACIFICA (*Coq.*).

Blatta pacifica, *Coqueb. Illustr. Ins.* i. p. 91, pl. xxi. f. 1 (1801).

11. SCAPTERISCUS ABBREVIATUS, *Scudd.*

Scapteriscus abbreviatus, *Scudd. Mem. Peabody Acad. Sci.* i. p. 14, t. i. ff. 8, 20 (1869).

Larvæ found in burrows in the sand under a *Conferva* (*Enteromorpha*), in salt water, on the shore of San Antonio Bay, a little above high water. Perfect insect in and about the yards of the house.

12. GRYLLUS ASSIMILIS (*Fabr.*).

Acheta assimilis, *Fabr. Syst. Ent.* p. 280, n. 3 (1775).

A species widely distributed in America. This is the black cricket mentioned by Webster and other visitors to the island. It is very common in the central district on the paths, and makes a great noise, especially about 4 o'clock in the afternoon.

13. GRYLLUS FORTICEPS, *Sauss.*

Gryllus forticeps, *Sauss. Miss. Sci. Mex.* vi. p. 407 (1870).

14. ŒCANTHUS (?) PALLIDOCINCTUS, n. sp.

Long. corp. 13 millim.

Male. Reddish brown, abdomen, hind knees, and hind tarsi darker; head rather flattened, the palpi and the outside of the scape of the long and slender antennæ towards the base whitish. Pronotum longer than broad, sides subparallel, more shining and paler on the lower lateral border than above; elytra brown, reticulated with darker nervures, about two fifths as long as the abdomen, which is rather long and cylindrical, the incisions conspicuously pale; cerci broken, but the remaining portions are bordered on each side with very long and fine hairs. Legs short and stout, and slightly compressed; all the femora enlarged, the middle ones least so; hind tibiæ spined from the base, with 3 large terminal spines on each side, and 3 or 4 larger spines alternating with smaller ones beyond the middle; first joint of tarsi with 2 small and 1 large spine on each side.

Probably belongs to a new genus of *Œcanthidæ*, but has a superficial resemblance to *Gryllodes*.

15. CONOCEPHALUS VERNALIS, n. sp.

Exp. al. circa 78 millim.

Bright grass-green; wings hyaline, with bright green nervures; fastigium short, obtusely rounded, as in *C. triops*, Linn.; four front femora unarmed. Male with face, antennæ, eyes, and the whole of the fastigium whitish; tips of mandibles and palpi red; tibiæ paler than the ground-colour; auditory apparatus and tarsi beneath blackish; tegmina yellowish along the costa, a buff streak at the base above the subcostal nervure, and a white basal streak on the left side above the median nervure. Female much less varied with white; labrum white, mandibles more yellowish, auditory apparatus and tarsi whitish, the latter brown below; ovipositor a little longer than the abdomen, extending far beyond the knees, but shorter than the closed wings; ovipositor paler towards the extremity, and slightly veined with reddish, the extreme tips of the blades black.

This grasshopper was very common everywhere in the Main and Rat Island. It makes a great noise at night. The brown form, which flew about with it, was not so common.

Var. FRATER.

Exp. al. 75-83 millim.

Brown; the thorax apparently rather longer and narrower, an effect which is caused by an indistinct pale or blackish line on each side; frequently a pale streak behind each eye; mandibles varied with pale reddish, the extreme tips black; tegmina greyish brown, often with dark speckles above and below the principal nervures and on the costa beyond the middle (similar speckles are visible on one of the green specimens); wings hyaline, with greyish-brown nervures.

Genus ŒCELLA, n. gen.

Affinities doubtful; nearest to *Bargilis* in the structure of the legs, but more nearly approaching *Elimæa* in neuration.

Vertex slightly convex, shaped nearly like an equilateral triangle when viewed from above, the fastigium projecting about as far as the scapes of the antennæ, when the latter are recurved. Eyes large, round, prominent. Antennæ slender, filiform, at least as long as the tegmina. Pronotum above flat, oblong, slightly narrowed in front and rounded behind, excised laterally at the base of the wings. Tegmina rather narrow, especially at the tips, but hardly pointed; shorter than the wings, the drums of the male triangular, vitreous, nearly alike on both tegmina. Front legs moderately stout, unarmed; the tibiæ much swollen at the base to receive the linear foramina, which are well marked on both sides. Tarsi of all the legs similar, of nearly equal size, and about as long as broad, the second joint lobate, the last very slender, and nearly as long as the two preceding joints together. Middle tibiæ with a few very short and small spines towards the extremity. Hind legs long and slender, the basal half of the femora moderately thickened, furrowed on the side, and carinated below; tibiæ with a double row of short and slender spines, and with two small spines at the tip. Cerci of the male nearly as long as the short abdomen, with long terminal forks, the longest slender and almost sickle-shaped at the tip; subgenital laminæ short, concave at the extremity. Ovipositor nearly as long as the abdomen, moderately broad, strongly compressed, turned upwards, and pointed at the tip.

16. ŒCELLA FURCIFERA, n. sp.

Exp. tegm. 38 millim.; exp. al. 42 millim.

Head, thorax, tegmina, tips and veins of wings, and greater part of legs grass-green: antennæ and tibiæ, abdomen and appendages yellowish, a yellowish line on each side of the pronotum above; wings hyaline.

Hab. Pernambuco and Fernando Noronha. In all the specimens from the latter locality the green colour has more or less faded to yellowish brown. When fresh this insect is bright green. It was common on Main Island, and especially on Rat Island.

17. MERONCIDIUS VIRIDINERVIS, n. sp.

Exp. al. 65 millim.; long. corp. 38 millim.

Male. Brown, head smooth; labrum greenish; mandibles black, except at the base; a depressed circle surrounding the space of the antennæ, and the fastigium projecting in a spoon-shape between them; scape pointed; flagellum broken: the basal joints varied with lighter and darker brown; thorax strongly granulated, a little speckled with black, and much raised behind, where it assumes a slight greenish tint; tegmina brown, minutely reticulated and spotted with dark brown, chiefly above and below the nervures; longitudinal nervures mostly green; in the costal area the nervures are blackish towards the base, where they anastomose a little; on the disk the transverse nervures are brown or indistinctly green; inner margin with alternate darker and paler spaces: wings smoky hyaline, with reddish-brown longitudinal and brown transverse nervures; hind margins damaged, but probably browner than the rest of the wing; legs indistinctly mottled; spines of femora mostly black on the inner sides, hind femora with a black basal streak on the outside.

Somewhat resembles *M. indistinctus*, Walk., but the wings are shorter.

A single specimen on a tree in the Sapate.

18. STENOPOLA DORSALIS (*Thunb.*).

Truxalis dorsalis, *Thunb. Nov. Acta Upsal.* ix. p. 80 (1827).
Stenopola dorsalis, *Stål, Recensio Orth.* i. p. 83 (1873).

The hind legs have not been described; they are reddish brown, the middle of the femora being black on both surfaces, the striations more or less marked with paler. The hind tibiæ are armed, except on the basal third, with a double row of moderately long and pointed spines, the intermediate space above is clothed with long fine white hairs, and there is a row of much shorter white hairs on the under surface also. The sides and under surface of the hind tibiæ are generally dark green or blackish; at the tip there are two short spines on the outside, and two long ones on the inside. There are apparently only three joints to the hind tarsi: the first is three times as long as broad, but is broad and flattened; the second is much narrower, half as long again as broad, and produced into a long tooth at the extremity beneath, and the terminal joint is very slender at the base, gradually enlarging to the pulvillus, and about as long as the two preceding joints together.

The species appears to be common, and differs considerably in size, the tegmina expanding from 30–40 millim., and the body measuring from 18–20 millim. in length.

On both Main and Rat islands, but especially common on the latter.

19. PANTALA FLAVESCENS (*Fabr.*).

Libellula flavescens, *Fabr. Ent. Syst. Suppl.* p. 285 (1798).

Distribution. World-wide.

Very common everywhere on Main Island. The larvæ living in puddles in the central district.

20. TRAMEA BASALIS (*Burm.*).

Libellula basalis, *Burm. Handb. Ent.* ii. p. 852, n. 25 (1839).

Less common than the preceding. The abdomen when fresh is dark crimson-red. A well-known South-American species.

21. BLASTOPHAGA OBSCURA, n. sp.

Male. Long. corp. 2 millim.

Brown or yellowish brown, smooth, except a few short hairs on the tarsi. Front tarsi apparently 3-jointed, middle and hind tarsi 5-jointed; tarsal claws very strong, and front and hind tibiæ ending in short strong spines.

This species resembles the description of *B. brasiliensis*, Mayr, from Blumenau, but is considerably larger. I think it useless to give a detailed description, for which a larger series, including both sexes, and preserved in different ways, would be desirable. The locality will probably serve to fix the species, especially as true *Blastophaga* does not appear to be well represented in America.

22. GANOSOMA DISPAR, n. sp.

Male. Long. corp. 1½ millim.

Yellow, smooth; head forming a long oval, broad behind, gradually narrowed in front, antennæ inserted widely apart; legs of nearly equal size and structure, femora slightly thickened; tibiæ spinose on the outer edge, and terminating in a coronet of short spines, none of which are conspicuously longer than the rest; first joint of the tarsi longer than thick; abdomen long, tapering.

Differs from *Ganosoma attenuatum*, Mayr (♂), in not being depressed, and (perhaps) in the long abdomen ; and from *Tetragonaspis gracilicornis*, Mayr (♀), in the much shorter joints of the antennæ.

Female. Long. corp. 2 millim.; ovipositor $4\frac{1}{2}$ millim.

Tawny-yellow, with a slight greenish-coppery reflexion (colours perhaps altered by spirit) ; antennæ 12-jointed, serrated and set with very short hairs ; brown, except the two basal joints; scape as long as the three following joints, second joint rather longer than the fourth, third (annulus) very small, fifth and following gradually smaller, the last three joints forming a club ; ovipositor more than twice as long as the body ; veins of the wings of nearly uniform thickness ; ulna as long as the pterostigma, hardly curved or thickened, metacarpus about as long as the radius. Head and thorax finely rugose. Legs yellow, the femora slightly thickened.

Appears to approach most nearly *Tetragonaspis* * *flavicollis*, Mayr, but that species has two annuli (ring-joints) to the antennæ. Except in the structure of the antennæ, the single specimen before me much resembles the figure of *T. gracilicornis*, Mayr, but the latter species has longer hairs on the antennæ, and the terminal joints do not form a club, to say nothing of other differences.

23. EVANIA LÆVIGATA, *Latr.*

Evania lævigata, *Latr. Gen. Crust. Ins.* iii. p. 251 (1807).

A cosmopolitan species. The larvæ of this genus are parasitic in the egg-capsules of Blattidæ.

A single specimen taken in a house at Sambaquichaba. It is very common in Pernambuco.

24. CAMPONOTUS BIMACULATUS (*Smith*).

Formica bimaculata, *Smith, Cat. Hym. B. M.* vi. p. 50, n. 171 (1858).

Six specimens, all small workers. The species is new to the Museum collection. Smith described it from St. Vincent's. Roger (Berl. ent. Zeitschr. vi. p. 285, 1862) identifies this species with *Formica ruficeps*, Fabr. (Syst. Piez. p. 404, n. 32) ; but I am not satisfied that this is correct, as Fabricius does not mention the conspicuous pale spots on the second segment of the abdomen. In the small workers the head is mostly black;

---

* This genus proved to be the female of *Ganosoma*.

in the large workers it is red. Dr. Mayr records this species from New Granada, and it is probably widely distributed in South America.

Under the wood of a Burra tree in the Sapate.

25. PHEIDOLE OMNIVORA, n. sp.

Soldier $4\frac{1}{2}$ millim.; worker $2\frac{1}{2}$–3 millim. in length.

Soldier red, shining (abdomen darker), sparingly covered with raised white hairs. Head, without the mandibles, about as broad as long, finely and sparingly longitudinally striated above and in the middle, where it is depressed, behind; sides gradually rounded behind the eyes. Mandibles very broad, hardly punctured, blackish at the base and tips. Scape of the antennæ straight, gradually thickened beyond the middle, nearly one third of the length of the antennæ. Second joint about three times as long as broad, joints 3–9 very slightly longer than broad, joint 10 much thicker and twice as long as broad, joint 11 rather shorter and thicker, and joint 12 forming a long pointed cone; the hair on the antennæ is thicker and closer than on any other part of the body. Mesonotum with a hump on each side above, and finely punctured above; metanotum, which is armed with two strong spines, more closely; first node of the petiole with the spine somewhat truncated at the extremity; second node fully twice as broad as the first, and with strongly projecting lateral angles; abdomen finely punctured at the base.

Worker similar, but smaller and generally darker, the antennæ lighter; the scape nearly as long as the remainder of the antennæ, slightly curved, but hardly thickened, and the terminal joint thicker in proportion and less pointed than in the soldier. The spines on the metanotum and on the first node of the petiole are much shorter, and the projecting sides of the second node are rounded off.

This species much resembles *Pheidole pusilla*, Heer, in size and general appearance, but differs widely in structure.

This was exceedingly abundant in the houses, making nests in the earth between the bricks of the floor. It is very destructive. devouring all kinds of food, and even ate up the insects we captured, in the chip-boxes.

26. MONEDULA SIGNATA (*Linn.*).

Vespa signata, *Linn. Syst. Nat.* ed. x. i. p. 574, n. 14 (1758).

A common South-American species.

Very common on the sand-hills, where it makes its burrow.

27. POMPILUS NESOPHILUS, n. sp.

Long. corp. 10 millim. ; exp. al. 15 millim.

Female. Dull black, first three segments of abdomen, the sides of the fourth, and the under surface of the hind tibiæ red. Wings smoky, a little lighter and subhyaline on the hind wings and towards the base of the fore wings.

Head large, eyes hardly extending to the base of the jaws, face rather more strongly punctured than the vertex ; clypeus short, slightly emarginate ; labrum short, transverse ; eyes nearly parallel, front ocellus forming the apex of a rectangle with the hinder ocelli, the two latter rather wider apart than the space between these ocelli and the eyes ; second joint of the antennæ half as long again as the third. Pronotum falling in front to the neck in a short rounded curve, rather broader than the meso- or metanotum ; the metanotum is furnished with a small prominence on each side near the base.

Second and third cubital cells of nearly equal size, the second recurrent nervure striking the third cubital cell in the middle.

Spines of the legs as usual.

Much resembles *P. sobrinus*, Blanch., a Chilian species, in which, however, the thorax is verdigris-green above, instead of black.

Taken flying over paths in the centre of the island, not common and difficult to catch.

28. TACHYTES INCONSPICUUS, n. sp.

Long. corp. 6–8 millim.

Black, clothed with a very fine silvery pile (that on the upper part of the face with a slight golden appearance in certain lights), otherwise most conspicuous on the sides of the abdomen, towards the extremities of the segments ; thorax and scutellum shining, with very numerous small punctures, not very close together : metathorax rugosely-punctate ; tegulæ testaceous ; wings clear hyaline, with a strong violet iridescence ; nervures dark brown.

Closely allied to *T. iridipennis*, Smith, from Ega, but in that species the thorax is longitudinally striated, and the tips of the tarsi are ferruginous.

29. Polistes Ridleyi, n. sp.

Long. corp. 15–16 millim.; exp. al. 26–20 millim.

Worker. Varied with ferruginous red, dark brown, black, and yellow; clothed with very fine silky pubescence. Head red, the vertex darker, the face and orbits lighter, base of the head black; antennæ black in the middle above. Thorax black, the prothorax and mesothorax red above, the hinder edge of the prothorax paler, and the front edge narrowly bordered with yellow; the edges and central line of the mesothorax more or less bordered with black above; on the pleura, beneath the fore wings, is a conspicuous yellow spot; scutellum and post-scutellum edged in front with yellow, the band on the former hardly complete in the middle, and the space behind it more or less red; metathorax finely and transversely striated, a deep longitudinal channel in the middle, more strongly striated, and edged with a yellow stripe on each side; on each side, above the base of the hind coxæ, is another yellow spot. Abdomen dark brown, finely pubescent, shading into black at the base, and generally more or less red at the extremity. Legs red, coxæ and femora black, knees red or yellow, hind tibiæ more or less black in the middle. Wings smoky hyaline, strongly tinged with ferruginous along the costa of the fore wings; tegulæ ferruginous.

Var. *a.* Face, head beneath, pectus, and coxæ beneath yellow; femora striped beneath with yellow; first two segments of the abdomen with a small yellow spot on each side above, and a larger one near the base of the first segment beneath.

Very nearly allied to *P. instabilis*, Sauss., from Mexico; but this is a redder insect, with the segments of the abdomen always more or less bordered with yellow.

This insect is called here "*Marimboudo*," and is very common. It makes its nest on the underside of an overhanging rock or eaves of a house, or on the branch of a tree. I have seen a Cashew-nut tree containing an immense number of nests in various stages of construction. The nest consists of a single comb of cells of a triangular or oval outline, and attached by a pedicel at the narrow end; a large one is about four inches in length, and three across in the broadest part. The cells are about three quarters of an inch deep, and a quarter of an inch across. The insect stings slightly, but only when much irritated.

It plays a very important part in the fertilization of the flowers, especially the Cucurbitaceæ.

30. HALICTUS LÆVIPYGA, n. sp.

Female. Long. corp. 10 millim.: exp. al. 16 millim.

Head and thorax dark green, slightly bronzed; abdomen shining, shading into violet at the extremity of the segments; wings hyaline; legs black, clothed with pale hairs.

Head and face finely and closely punctured, sparingly studded with whitish hairs, dark green, occasionally shading into bronzy or violet in certain lights; clypeus green or violet-black, with much larger and fewer punctures than the upper part of the head; the extremity and the labrum black, the lower mouth-parts dull bronzy green; antennæ black, ferruginous towards the extremity beneath: thorax dark green, thickly punctured, most finely on the prothorax, which shades into bronzy; metanotum and base of abdomen above densely clothed with whitish hairs; metanotum with longitudinally diverging striæ in front, behind deeply and longitudinally concave, with small punctures, the intermediate spaces still more finely punctured; tegulæ testaceous; wings hyaline, with brown nervures and a slight violet iridescence; pectus dull green, densely clothed with whitish hairs; propectus violet-black: legs black, with whitish hairs above, and the tibiæ and tarsi densely clothed with red hairs beneath; the knees (very narrowly), spines, and last joint of the tarsi mostly reddish. Abdomen smooth and shining, the punctuation being extremely fine, even under the microscope, and densely clothed with short white hairs, except towards the extremity of the first two segments; the segments greenish in front and violet-black behind; towards the extremity the lateral bristles are longer; under surface of abdomen brown, the extremities of the segments slightly reddish.

31. HALICTUS ALTERNIPES, n. sp.

♂ ♀. Long. corp. 11 millim.; exp. al. 13–14 millim.

Head and thorax dark green above; abdomen green, smooth and shining, the neighbourhood of the incisions more or less violet; wings hyaline, or slightly clouded in and below the radial cell, and towards the extremity of the discoidal cell of the hind wings; legs reddish, the femora and tibiæ in the middle, and often the last joint of the tarsi dark green or blackish: clypeus tipped with tawny, and with more numerous punctures than in *H. lævipyga*; the punctures of the mesothorax and the lower part of the metathorax larger; the channel of the metathorax narrower; antennæ black above, ferruginous below; cheeks bronzed.

Closely allied to *H. lævipyga*, but differs · in colour and punctuation. Appears to be a very common species.

32. HALICTUS ATRIPYGA, n. sp.

♂ ♀. Long. corp. 5 millim.

Head and thorax clothed with whitish hairs, green, very finely punctured; head much bronzed; clypeus rather short, the tip black; metanotum longitudinally striated in front; abdomen bronzy black, more violet in the incisions; antennæ black, ferruginous beneath; head and thorax dull bronzy green beneath; abdomen beneath shading more into ferruginous brown; wings hyaline, with brown nervures; tegulæ pitchy; legs ferruginous, the femora, and sometimes more or less of the tibiæ, blackish.

A small species, not closely allied to any other, and easily recognizable by the strong bronzy colouring of the back of the head.

These *Halicti* were taken in the flowers of the Melons and the *Oxalis*, and play an important part in the fertilization of the flowers in the island.

33. TARUCUS HANNO, *Stoll.*

Papilio Hanno, *Stoll, Suppl. Cram.* pl. xxxix. ff. 2, 2 B (1790).
Rusticus adolescens Hanno, *Hübn. Samml. ex. Schmett.* i.

Many specimens, taken between July 25 and August 31.

The Brazilian form of this widely distributed and variable species. It is generally larger than the more northern form of the species, and stands in the British-Museum collection under the MS. name of *T. monops*, Zell. The specimens expand from $17\frac{1}{2}$ to 23 millim.

This little Blue was very common on Rat and Main islands. The *Catachrysops trifracta*, Butl., said to have been caught on Rat Island by the 'Challenger' Expedition, we did not meet with; and it seems possible that there was some mistake in the labelling of this specimen, as the genus is only known from the East Indies.

34. HELIOTHIS ARMIGER (*Hübn.*).

Noctua armigera, *Hübn. Eur. Schmett.*, *Noct.* f. 370.

A cosmopolitan species.

A single specimen was caught flying in the daytime on East Hills.

35. ANOMIS (?) DISPARTITA, *Walk.*

Anomis (?) dispartita, *Walk. Cat. Lep. Het.* xiii. p. 990, n. 8 (1857).

Described by Walker from Jamaica and St. Domingo; the specimens from Fernando Noronha most nearly resemble that obtained in the former locality.

Taken on the wing at night. Main Island.

36. ANTHOPHILA FLAMMICINCTA, *Walk.*

Anthophila flammicincta, *Walk. Cat. Lep. Het.* xxxii. p. 801 (1865).

The types are from St. Domingo.

37. BOLINA BIVITTATA, *Walk.*

Bolina bivittata, *Walk. Cat. Lep. Het.* xiii. p. 1156, n. 23 (1857).

A common species. The specimens in the British Museum are from Honduras, Venezuela, St. Domingo, and Jamaica. The single specimen from Fernando Noronha has a narrower band than any of the others; but it would require a series to show whether this character is constant in the locality, or a mere individual variation.

38. THERMESIA GEMMATALIS (*Hübn.*).

Antisarsia gemmatalis, *Hübn. Zutr. ex. Schmett.* i. p. 26, n. 77, ff. 153, 154 (1818).

Common among the wild beans on Main and Rat Islands (August 17–25).

39. NEMORIA DENTICULARIA, *Walk.*

Nemoria (?) denticularia, *Walk. Cat. Lep. Het.* xxii. p. 536, n. 20 (1861).

The locality of the type specimen is unknown; but it is probably a wide-ranging species, as there are specimens in the British Museum from Corrientes and Goya.

40. ACIDALIA FARA, n. sp.

Exp. al. 16 millim.

Male. Grey, thickly dusted with brown, the first line commencing at about one third of the length of the wing with a dark shade curving to a conspicuous black spot, thence curving sharply inwardly till it terminates in a more conspicuous dark acute angle projecting outwardly just above the inner margin; it is not continued on the hind wings. The second line commences at two thirds of the length of the wing in another dark shade on the costa, which runs obliquely to another black spot, below which it is continued in grey festoons across both pairs of wings, forming more than a half-circle on the basal side of a conspicuous black spot on the hind wings. The third line is the most conspicuous, and is formed of a series of black spots connected by a grey line on the fore wings, which curves inwards twice, a little above the middle, and again above the inner margin. On the hind wings it forms a nearly continuous black line, curving outwards at one fourth of the distance from the costa, and again, in a wider and more regular curve, below the middle. At the base of the fringes runs a row of black dashes, and between these and the third line are a series of indistinct greyish markings. Underside grey-white, with the central spots indicated, but not conspicuous. Head black above; body and legs grey; abdomen rather indistinctly banded with black above, and with a blackish space at the base of the anal tuft; femora tawny beneath, front femora and tibiæ blackish above: antennæ tawny, alternating with whitish above.

Allied to *A. retractaria* from Florida, but darker and more distinctly marked.

41. PYRALIS MANIHOTALIS, *Guén.*

Pyralis manihotalis, *Guén. Spéc. Gén. et Lép., Delt. et Pyr.* p. 121 (1854).

Described by Guénée from Cayenne.

42. SAMEA CASTELLALIS, *Guén.*

Samea castellalis, *Guén. Spéc. Gén. Lép., Delt. et Pyral.* p. 195 (1854).

A widely distributed species throughout America.

43. HYMENIA PERSPECTALIS (*Hübn.*).

Pyralis perspectalis, *Hübn. Eur. Schmett., Pyr.* f. 101.

A widely-distributed species.

44. PHAKELLURA HYALINATA (*Linn.*).

Phalæna-Geometra hyalinata, *Linn. Syst. Nat.* i. (2) p. 874, n. 279 (1767).

This little moth was very common on the Main Island, especially about the Peak, among the wild melons, &c. (August 17–20).

A cosmopolitan species.

45. MARGARONIA JAIRUSALIS, *Walk.*

Margaronia jairusalis, *Walk. Cat. Lep. Het.* xviii. p. 524, n. 15 (1859).

Originally described from Venezuela.

46. ACHARANA PHÆOPTERALIS (*Guén.*).

Botys phæopteralis, *Guén. Spéc. Gén. Lép., Delt. et Pyr.* p. 349 (1854).

A common and widely-distributed species in Tropical America.

47. PACHYZANCLA DETRITALIS (*Guén.*).

Botys detritalis, *Guén. Spéc.Gén. Lép., Delt. & Pyr.* p. 347 (1854).

Widely distributed in South America.

48. OPSIBOTYS FLAVIDISSIMALIS (*Grote*).

Botis flavidissimalis, *Grote, Canad. Ent.* ix. p. 105 (1877).

Described by Grote from Texas ; but there are also specimens from South America in the British Museum.

49. MELLA ZINCKENELLA (*Treitschke*).

Phycis zinckenella, *Treitschke, Schmett. Eur.* ix. (1) p. 201 (1832).

Many specimens. A cosmopolitan species. There are specimens in the British Museum from S. Europe, S. Africa, and Australia. It is a somewhat variable insect, but is easily recognized by the broad pale costa and the huge palpi.

50. PENTATOMA TESTACEA, *Dall.*

Pentatoma testacea, *Dall. Cat. Hem.* p. 250, n. 43 (1851).

Several specimens were met with in various spots in the Main and Rat Islands.

51. LYGÆUS RUFOCULIS, n. sp.

Long. corp. 9-11 millim

Bright red, including the eyes and ocelli ; head above, between and in front of ocelli, blackish ; antennæ and legs blackish and setose ; coxæ reddish at base, corium more or less varied with blackish, outer edge narrowly black, membrane brown, blackish in ♀. Head and thorax rugose, the former truncate and bicarinate in front. Rostrum black, extending to the extremity of the posterior coxæ. Corium fully developed in ♂; in ♀ about half as long as the abdomen, and rather pointed.

An extremely common species, allied to *L. anticus*, Walk., in which, however, the head and thorax are unicolorous red, and the tegmina are reddish brown.

This highly-coloured Bug occurred in large numbers upon the ground in the Sapate under bushes of *Jacquinia* ; but was local, only found in a few spots.

52. HERÆUS VARIEGATUS, n. sp.

Long. corp. 7 millim.

Head and front of pronotum dull red ; posterior lobe of pronotum testaceous, with very large separated punctures. Corium blackish, with rows of punctures : outer edge of corium testaceous, with two black spots, one at its extremity, the other a little before ; the veins, a large triangular spot on the inside, and three smaller spots between this and the margins are also testaceous ; scutellum with a converging testaceous line on each side ; membrane blackish, with a testaceous blotch at base and tip. Antennæ finely pubescent, the two basal joints testaceous, the third blackish, the fourth black, with the base white. Legs testaceous, front and hind femora and all the tibiæ strongly spined. Body red beneath, pectus darker. Rostrum testaceous, extending to the base of the hind coxæ.

Allied to *H. percultus*, Dist.

Taken at light.

53. LIGYROCORIS BALTEATUS, *Stål.*

Ligyrocoris balteatus. *Stål, Vet.-Akad. Handl.* (2) xii. p. 146, no. 4 (1874).

Flew into light.

54. LIGYROCORIS BIPUNCTATUS, n. sp.

Long. corp. 4 millim.

Head, pronotum, and scutellum black ; hinder lobe of pronotum with two short, reddish, longitudinal stripes in the middle, a reddish spot on each side in front and one at each angle behind ; antennæ testaceous, the last joint black. Corium testaceous, with rows of black depressed punctures in the middle ; the veins on the inner edge black, running into a large irregular apical black border ; the outermost vein broadens out just before reaching it and encloses a very conspicuous oval white spot ; membrane greenish, with two brown, diverging, curving lines in the middle, two brown spots at the base, and one on each side of the curve. Under surface of the body apparently black ; legs testaceous, femora more or less black ; but the specimen is carded in such a manner as not to allow of a proper examination of the under surface or of the legs and rostrum.

55. RHAGOVELIA INCERTA, n. sp.

Long. corp. 2¼ millim., lat. 1 millim.

Blackish brown ; the orbits, front of prothorax, whitish, slightly tawny; abdomen with a whitish pile on the sides and less conspicuous beneath : base of antennæ, femora above, and legs beneath tawny : closed hemilytra whitish (or slightly tawny) at the base between the nervures, but with no other pale markings except the conspicuous long-oval white spot before the tip.

A shorter and broader species than any at present represented in the British Museum ; but with most general resemblance to *Microvelia vagans*, White. It is possibly a variety of the widely distributed and variable *Velia collaris*, Burm.

On grasses in the lake.

56. PSILOPUS METALLIFER, *Walk.*

Psilopus metallifer, *Walk. List Dipt. B. M.* iii. p. 647 (1849).

Flying over Cucurbitaceæ on the Peak.

57. TEMNOCERA VESICULOSA (*Fabr.*).

Syrphus vesiculosus, *Fabr. Syst. Antl.* p. 226, n. 11 (1805).

Flying over herbage in the sun, and also taken on the flowers of the cabbage in the garden.

58. SARCOPHAGA CALIDA, *Wiedem.*

Sarcophaga calida, *Wiedem. Aussereurop. zweifl. Ins.* ii. p. 366, n. 24 (1830); *Walk. Ins. Saund. Dipt.* p. 326 (1856).

COLEOPTERA.

By CHAS. O. WATERHOUSE, F.E.S.,
Assistant in the Zoological Department, British Museum.

The number of species obtained was 24. As might naturally be expected, many of them are Brazilian species or are species with a wide geographical range. One of the Heteromerous genera, which I have named *Æsthetus*, is almost certainly indigenous. Perhaps the most interesting species is a Longicorn of the genus *Acanthoderes*. At first sight I took it to be a pale variety of *A. jaspidea*, a common Brazilian species, but on closer examination the apex of the elytra was found to be different. If this species should hereafter prove to be peculiar to the island, it will be a somewhat remarkable fact.

A few of the species the determination of which would have been attended with great labour, I have left unnamed, as I feel sure they are introduced species.

BRACHELYTRA.

BELONUCHUS, sp.

A single specimen, closely resembling *B. formosus*, Grav., of Brazil, but smaller, 2¼ lines in length. It was found in a decomposing rat.

TROGOPHLŒUS, sp.

A single example, closely resembling *T. pusillus*, Grav., of Europe, but a trifle larger, with the two basal impressions of the thorax strongly marked.

NECROPHAGA.

DERMESTES FELINUS, *Fabr.*
A widely-distributed species.

EPURÆA ?
A single specimen of a very small species (1¾ millim.) somewhat resembling *Epuræa limbata*, but a little narrower, with the margins not expanded, light brown, finely pubescent, closely and very finely punctured; each elytron having two fine, widely separated, impressed striæ. Abdomen covered by the elytra.

PALPICORNIA.

PHILHYDRUS MARGINELLUS, *Fabr.*
A widely-distributed species.

DACTYLOSTERNUM ABDOMINALE, *Fabr.*
A widely distributed species.

LAMELLICORNIA.

ATÆNIUS, sp.
Three examples belonging to two species. No doubt introduced.

STRATEGUS ANTÆUS, *Fabr*.

The prothorax and elytra of this North and Central-American species. [Its larvæ were found under rubbish in the Sapate with the fragments of the perfect insect, which we did not succeed in taking alive.—H. N. R.]

SERRICORNIA.

HYPORRHAGUS MARGINATUS, *Fabr*.

A single example of this West-Indian species in the flowers of an acacia in the garden.

MALACODERMATA.

XYLOPERTHA, sp.

Three specimens. Probably an introduced species. Four millimetres in length; reddish yellow, shining, with the rough anterior part of the thorax brown and the apex of the elytra pitchy. The elytra finely punctured at the base, strongly punctured posteriorly; the posterior declivity also strongly punctured, with three nodes on its upper margin on each elytron.

[These came to light in our rooms.—H. N. R.]

HETEROMERA.

EPITRAGUS BATESII, *Mäklin*.

Elongatus, ellipticus, modice convexus, parum nitidus, parce flavo-pubescens, crebre punctatus; elytris striato-punctatis, striis postice obsoletis.

Long. 7½ millim.

The head is closely and rather strongly punctured. The thorax two fifths broader than long, broadest a little behind the middle, very slightly narrowed behind, a little more in front; distinctly and moderately strongly punctured; the punctures on the disk separated from each other by one or one and a half times the diameter of the punctures; the punctures towards the sides larger and closer together, giving a slight rough appearance; the punctures at the anterior margin are finer. The elytra are at the base a little wider than the base of the thorax, very slightly widened to rather behind the middle and then arcuately narrowed to the apex; moderately finely but distinctly punctured, the punctures not very close together; with several short lines of rather larger punctures, which are most distinct towards the margin. Antennæ and legs pitchy.

Numerous examples of this species from the Amazons are labelled in Mr. F. Bates's collection with the name " *E. Batesii*, Mäkl.," but the species does not appear to be described.

Blapstinus Ridleyi, n. sp.

Elongatus, oblongo-ovatus, parum nitidus, fusco-brunneus, flavo-pubescens; thorace crebre evidenter punctato; elytris punctato-striatis, interstitiis lateralibus et ad apicem convexiusculis, subtiliter vix crebre punctatis; antennis, tibiis tarsisque piceis.

Long. 5 millim.

Antennæ with the third joint elongate, about one fourth longer than the second, the fourth a trifle shorter than the third, the fifth, sixth, and seventh about as long as broad, the eighth, ninth, and tenth slightly transverse. The head is moderately strongly punctured, but the punctures are not *very* close together; the epistome is moderately emarginate, rather more closely punctured than the head. The thorax is evenly convex, broadest *at* posterior angles, scarcely sinuate at the sides, narrowed at the anterior third; moderately strongly punctured, the intervals between the punctures about equal to the diameter of the punctures; the anterior angles moderately prominent and acute; the base rather strongly bisinuate. The elytra a little broader than the thorax; somewhat strongly punctate-striate; the striæ near the suture scarcely impressed at the base; the punctures in the striæ moderately large and close together; the punctures on the interstices are fine but distinct, the spaces between them being about once and a half the diameter of the punctures.

I have ventured to describe this species as it appears to be new, although probably introduced.

Blapstinus, sp.

Several specimens of a species closely resembling *B. pulverulentus*, Esch., but with the striæ of the elytra more impressed. There are several North-American species closely allied to this with which I am unacquainted, and it is not improbable that it is referable to one of them.

ÆSTHETUS, n. gen.

General characters of *Cyrtosoma*. Mentum small, narrowed at the base, truncate in front; ligula somewhat round; labial palpi short and stout, the apical joint rather large, ovate. Labrum nearly twice as broad as long, nearly straight in front, the angles rounded. Head transversely impressed between the eyes. Epistome not separated from the forehead by a distinct line, considerably advanced in front of the insertion of the antennæ, obliquely narrowed in front. Thorax evenly convex, the sides gently arcuate. Scutellum very small, short triangular. Elytra oblong-ovate, very convex, but somewhat flattened dorsally; their epipleura very broad and flat, gradually narrowed to the apex of the elytra. Wings absent. Prosternal process considerably produced posteriorly, acuminate, horizontal. Mesosternum sloping, slightly concave. Metasternum very short; intermediate and posterior coxal cavities separated by a very narrow space. Antennæ moderately long and slender, the third to seventh joints elongate, the apical joints a trifle broader. All the tibiæ slightly curved.

ÆSTHETUS TUBERCULATUS, n. sp.

Piceo-niger, nitidus; capite crebre sat fortiter punctato, epistomo convexo, subtiliter punctulato; labro piceo-flavo; thorace creberrime punctato, latera versus tenuiter ruguloso; elytris opacis, fortiter striatis, striis impunctatis, interstitiis sat convexis, singulis serie tuberculorum minutorum instructis; antennis, palpis tarsisque piceis, tibiarum apice intus, tarsisque subtus fulvo-pilosis.

Long. 9–11 millim.

The antennæ are somewhat slender; the second joint scarcely longer than broad, the third three times as long as the second; the fourth to seventh joints elongate, each a trifle shorter and broader at its apex than the previous one; the eighth, ninth, and tenth joints pilose, broader and shorter than the seventh, but not transverse; the eleventh oval. The thorax is evenly convex, very gently arcuate at the base, emarginate in front, moderately rounded at the sides, finely margined all round (except at the middle of the front margin), the posterior angles are *very* slightly projecting, the anterior angles slightly obtuse. The punctures on the disk are close and distinct, at the base and at the sides they are very fine and obscure. On each side of the disk the surface is finely longitudinally rugulose, but this is very slight in the larger examples. The interstices of the elytra are closely and very finely punctured, each interstice having a line of rather closely placed, minute, shining tubercles. Epipleura of the elytra dull. Under flanks of the prothorax dull and closely longitudinally striated. Sterna and abdomen shining, finely punctured.

The two smaller examples have the thorax relatively narrower than in the larger examples, and the rugulose surface more marked and more extended. These differences are no doubt sexual.

These were found under stones and bark in the woods of the Sapate.

COPIDITA, sp.

Several examples of a species which I am unable to determine. Yellowish, usually with slight grey shade on the elytra. Length 6–7 millim.

Those captured were attracted by a light.

BRUCHIDA.

BRUCHUS POROSUS, *Sharp.*

Two imperfect male specimens, which may be referable to *B. porosus*, Sharp (Biol. Cent.-Amer., Coleopt. v. p. 490), from Guatemala, the type of which (unique) is a female. The brown colour is rather more suffused over the elytra, and the punctures are not quite so large. The pectinations of the antennæ are very long.

RHYNCHOPHORA.

SITOPHILUS ORYZÆ, *L.*

Introduced. [It is very destructive to the maize-grains, so that in the store-rooms the maize is covered with a thick layer of sand to prevent their attacks.—*H. N. R.*]

XYLOPHAGA.

TOMICUS?

Two immature specimens belonging to this or an allied genus. Very pale yellow; 1½ millim. in length.

PYCNARTHRUM? SETULOSUM, n. sp.

Oblongo-ovatum, brunneum, sat nitidum, convexum; thorace latitudine paulo breviore, convexo, postice paulo angustiore, ante medium oblique angustato, creberrime subtiliter punctulato; elytris thoracis basi perparum latioribus, subtiliter striato-punctatis, interstitiis parce subtilius punctatis, parce pubescentibus, seriatim squamulato-setosis; antennis pedibusque sordide testaceis.

Long. 2 millim.

Head distinctly visible from above; concave in front in one sex. Eyes coarsely granular, widely separated above, but very slightly separated below. Antennæ testaceous; funiculus 6-jointed (or *possibly* 7)*; the first large, subglobose; the following joints very short and transverse, gradually increasing in width; club large, 3-jointed, oval, pubescent. The thorax has a well-defined margin separating the under flanks. The surface (seen through a microscope) is finely coriaceous, moderately finely punctured, the intervals between the punctures about equal to the diameter of the punctures; sparsely pubescent, the hairs at the front margin slightly thickened. The striæ of the elytra are lightly impressed, but scarcely so on the disk; the punctures in the striæ moderately fine and close together, the punctures on the interstices rather smaller and moderately widely separated. Anterior tibiæ rather broad, with four or five small obtuse teeth on the outer side, and two larger ones, one at one third from the apex, the other apical. Tarsi slender.

This insect agrees in the majority of its characters with *Pycnarthrum gracile*, Eichh. (Mém. Soc. R. d. Sci. Liège, viii. 1878, p. 104). The anterior tibiæ are, however, evidently different: " tibiæ anteriores apice extus rotundatæ." The structure of the antennæ appears to be the same, but the club is ovate and not acuminate. The elytra are punctate-striate and not crenate-striate, and the punctures are round and not subquadrate, &c.

It appears to be related to *Cnesinus*, Horn, but the anterior coxæ are not so widely separated.

[It was bred from the bark of the endemic fig-tree, from a specimen out of the garden of the Residency.—*H. N. R.*]

PLATYPUS PARALLELUS, *F.*

Two examples of this Brazilian species.

---

* The joints after the first are so confused that even with the antenna mounted in balsam I am not quite certain of their number.

LONGICORNIA.

ACANTHODERES RIDLEYI, n. sp.

Latus, depressus, omnino albo-grisco-pubescens; elytris pone medium macula laterali inconspicua ornatis; apice mucro brevi instructo.

Long. 11½–16 millim.

Form and general appearance of *A. jaspicea*, Germ., but a little more depressed and with the elytra rather more obtuse at the apex. The colour is pale whitish grey, generally with some sandy-yellow shade on the base of the antennæ, disk of the thorax, and on parts of the elytra. The front of the head is paler, with some conspicuous black punctures, especially between the eyes. The thorax has the usual median raised line and slight swelling on each side of the disk; there is no black at the sides; there is a line of very distinct black punctures along the basal margin, and a similar (but less regular) line along the front margin, and there are some other punctures scattered over the surface. The elytra have the usual costa distinct, slightly sinuous as in *A. jaspidea*. On the shoulder a few small tubercles may be traced through the pubescence. Some examples have scarcely any trace of spots, but most have a not very conspicuous pale fuscous spot behind the middle near the side, and behind this there are generally numerous black punctures which are surrounded by a brown shade; usually a short oblique brown line may be seen at the apex of the costa. Apical mucro shorter and less acute than in *A. jaspidea*. Abdomen with a slight grey shade in the middle, and a line of black dots on each side. Tibiæ unicolorous, or with a very slight pale brown spot near the apex. Some examples have a slight oblique brown spot on the disk, rather before the middle.

[These flew into light in the evening, and were very plentiful. They made a loud squeaking noise when caught.—*H. N. R.*]

TRYPANIDIUS ISOLATUS, n. sp.

Dense pallide grisco-pubescens; thorace guttis minutis nonnullis ornato; elytris nigro-punctatis.

Long. 13–15 millim.

Closely allied to *T. dimidiatus*, Th., but relatively a little narrower and quite differently coloured. The pubescence is very

42*

pale grey, some parts being a trifle paler than others; and there is a slight mixture of pale yellowish-brown pubescence, especially on the elytra. On the underside the pubescence is more sandy yellow, leaving the middle of the sterna and abdomen dark. The thorax has the usual line of large punctures at the base; a very slight raised line behind the middle of the disk; the lateral tooth small and acute. The elytra have the very slight costa near the suture a little less raised at the base than in *T. dimidiatus*, and the apical truncature of each elytron is not straight but has the angles rounded. The black punctures are arranged as in *T. dimidiatus*, but extend to the apex. There is a small pale spot at the base close to the scutellum, and a scarcely noticeable brown spot near the suture a little way from the base. One example has a slight whitish mark on the suture just before the middle, and a moderately broad whitish band near the apex, somewhat similar to the band in *T. dimidiatus*, but broader nearer the apex and less angular near the suture, where it is only carried up to a level with its origin on the margin.

This species is, in many respects, intermediate between *T. dimidiatus* and *T. melancholicus*.

[These also came to light with the preceding, but were rarer, only two being taken.—*H. N. R.*]

PSEUDOTRIMERA.

SCYMNUS, sp.

Two examples of a species resembling the Indian *S. xeranipelinus*, Muls., but a trifle smaller ($1\frac{1}{2}$ millim.) : uniform brownish yellow, with golden pubescence; punctuation of the elytra close, fine but distinct; metasternum very strongly punctured.

THYSANURA and COLLEMBOLA.

By H. N. RIDLEY, M.A., F.L.S.

IAPYX SAUSSURII, *Humbert, Rev. et Mag. Zool.* Sept. 1868, p. 351, pl. ii. figs. 1–5.

A single specimen of what I believe to be a young example of this little animal was obtained under a stone in the Sapato woods.

It corresponds closely to the figure given by Humbert, excepting that it is only 10 millim. in length instead of 22 millim., and that the antennæ are as hispid as those of *I. solifuga*, Halliday. In Humbert's figure the antennæ are quite glabrous, but as he does not mention this among the differences between his species and *I. solifuga*, it is possibly an error of the draughtsman. The number of joints in the antennæ is almost equal to that of *I. Saussurii*, and more than that of *I. solifuga*; but the animal is but little larger than the type specimens of *I. solifuga* in the British Museum. Another point of difference, though very slight, is in the forceps. In both the above quoted species there is a secondary tooth on the inner edge of each chela, besides several smaller rounded papillæ; now in the specimen from Fernando Noronha this secondary tooth is nearer the apex of the chela than is either of the others.

Distribution. The distribution of the whole genus is very little known at present, as specimens are not often collected. *I. solifuga* occurs in South Europe, Algeria, and Madeira (the var. *Wollastoni*). *I. Saussurii* was obtained in Mexico at Orizaba. Species are also recorded, but not described, from the United States and Calcutta.

Dr. Grassi, in p. 1 of "Progenitori degli Insetti e dei Miriapodi," gives *I. Saussurii* as from Brazil; but does not say whether he has seen Brazilian specimens; and in his list of species mentions it merely from Mexico, evidently using "Brasili" as a synonym for Mexico. It is more than probable that, if sought for, it will be found to occur also on the mainland of Brazil.

LEPISMA LEAI, n. sp.

Though it was to be expected that house-inhabiting Lepismas would occur here, I sought for them in vain until just as we were leaving, when a single large *Lepisma*, perhaps disturbed by the packing-operations, appeared. The specimen was somewhat damaged in capturing it; but as it seems to be undescribed, and is a very curious animal, I describe it as it is.

Corpus 19 millim. longum (setis exclusis), griseum. Caput parvum, subrotundatum. Antennæ filiformes (fractæ). Oculi minimi, nigri, pone basin antennarum positi. Prothorax 3 mm. longus, margine superiore recto, inferiore excavato, marginibus lateralibus productis; mesothorax et metathorax similes sed breviores. Pedes coxis valde crassis, brevibus, oblongo-ovalibus,

subtus duabus setis longis ; articulis secundis longioribus multo
tenuioribus pubescentibus, supra spina crassa armatis ; tertiis
tenuioribus, æquilongis, subtus setiferis ; tarsis multo brevioribus
setiferis, uncis parvis duobus terminalis.   Inter coxas tres
squamæ, ovales, obtusæ, quam coxas minores.   Abdominis
segmenta subæqualia, glabra.   Segmentum ultimum breve.   In
medio penultimi segmenti appendices duæ breves, complanatæ,
acutæ ; post eas duæ laterales, breves, teretes, hispidæ, tunc
duæ longæ multo longiores et tenues hispidæ, tunc duæ longæ
graciles hispidæ, et in medio appendix longissima unica, crassior,
setosa, articulata et annulata.

The abdomen was filled with some bright green substance,
which was emitted from the mouth when touched—apparently
green paint nibbled off the shutters.

The most nearly allied species to this which I have seen was
obtained in Socotra, and is now in the British Museum.   The
breadth of the thorax is greater than in *L. saccharina*, but the
head is not concealed as in some species.

LEPISMA CORTICOLA, n. sp.

Parva, 1 cm. longa, angusta, metallica plumbea, dorso arcuato
nec complanato.   Caput parvum, rotundatum ; oculi ad basin antennarum, rubri.   Antennæ graciles, annulatæ, hispidæ ; articulus
basalis maximus, reliqui breves, plurimi, crassiusculi.   Palpi
maxillares 5-articulati, articulis breviusculis.   Palpi labiales
breves, clavati ; articulus basalis brevis sectus, secundus longior,
tertius brevissimus coniens, quartus rotundatus brevis.   Thorax
angustus, quam abdomen vix latior.   Prothorax quam mesothorax longior, metathorax brevior, marginibus omnium ciliis
rigidis munitis.   Pedes longiusculi, hispidi ; coxæ breves, latæ
nec crassæ ; secundo paullo angustior et brevior ; tarsi longiusculi, triunguiculati ; squama inter coxas prothoracis ovata acuta
magna, alteræ minores.   Abdomen breviusculum ; appendices
segmenti ultimi graciles, hispidæ, brevin-culæ, subæquales.

In rotten wood and under stones in the Sapate and the base
of the Peak.   It also occurred on the mainland at Pernambuco
in similar localities.

The chief peculiarity of this *Lepisma* is its very rounded back,
resembling that of a *Machilis* rather than that of a typical
Lepisma.   It is a small active species occurring singly, of a dark
leaden-grey colour.   The scales resemble those of *L. saccharina*
in outline, but are more notched at the upper edge, and seem

also to have more numerous ridges. The thorax is not much broader than the abdomen, and the margin does not extend much beyond the feet.

MACHILIS, sp.

A single specimen of a very small brown *Machilis* was taken under a stone at the base of the Peak : but, by an accident, the specimen was destroyed, and we met with no others.

SEIRA MUSARUM, n. sp.

Minuta, gregaria, in vita metallica, 1 mm. longa. Caput rotundatum, hispidum ; oculi in maculis nigris fascia obscura sæpe connectente. Antennæ breviusculæ, violaceæ, hispidæ; articulo basali brevissimo, secundo et tertio subæquali basali duplo longioribus; articulo quarto triunciali, longissimo. Collum distinctum. Segmentum secundum corporis (prothorax) latum, marginibus rotundatum ; tertium brevius, quartum multo brevius, quintum longius, sextum quinto subæquale, septimum sexto triplo longius, terminalia brevia. Corpus in speciem ferme glabrum, insquamosum, flavescens, segmentorum basibus et marginibus cæruleo-purpureis. Pedes hispidi, primi breviusculi, secundi longiores, tertii longissimi, graciles.

Very abundant between the wet bases of the petioles of the bananas, at the base of the Peak.

It is very nearly allied to *S. Buskii,* Lubb., which was described from specimens found in a hot-house in England, and probably introduced with tropical plants. It differs in the absence of hairs round the neck and on the body, longer hind legs, and also in coloration. The spring resembles that of *S. Buskii,* and is rather hispid. The neck is very distinct. These small Collembola have been much neglected by collectors, and it is most probable that this species was introduced in the bananas.

ECHINODERMATA *.

There were not many species of Echinoderms found on the island, but the following were obtained:—

CIDARIS TRIBULOIDES, *Lamk.*

Very plentiful on the north side of the island in coral-reef pools, near Sambaquiehaba and Morro do Chapeo.

* The species were determined for me by Prof. Jeffrey Bell.

DIADEMA SAXATILE, *L.*

Two specimens from pools at Morro do Chapeo.

TRIPNEUSTES ESCULENTUS, *Leske.*

Very plentiful in rock-pools in Sponge Bay, sometimes almost filling a small pool.

OPHIURA CINEREA, *M. & Tr.*

Common under stones, north side.

OPHIACANTHA sp.

A very small specimen with the preceding.

OPHIOCOMA PUMILA, *Lütk.*

Young specimens.

OPHIOCOMA ECHINATA, *Ag.*

A single specimen from Portuguese Bay. Quite unknown to our guide, so it is probably rare here.

OLIGOCHÆTA.

By W. BLAXLAND BENHAM, D.Sc.

On February 2, 1889, I received a tube of small worms, which had been collected by Mr. H. N. Ridley in the island of Fernando Noronha, with the request that I would identify them. I gladly undertook to do so, and obtained permission to open the worms, if necessary; for it is now admitted that in most cases it is almost impossible to pronounce with certainty on the genus of an Earthworm from external characters alone; and although in some cases external characters may point to some particular genus, yet it is not always safe to rest content with such an indication, and we must examine the internal anatomy in order to be sure of the point.

The tube which I received contained six small, ill-preserved worms, one of which was a Polychæte, which I did not further examine. Of the remaining five, the first (which I will call A) was of rather an earthy-brown colour and measured 4 inches in length; the second and third were similar in colour and general appearance, but were only about 2 inches long; the fourth (B) was of a much darker tint, and was reddish brown in colour, somewhat like that of *Lumbricus terrestris* ; it measured 5½ inches ; the fifth turned out to be merely the anterior portion of a similar specimen.

The Worm A.—The first feature which struck me was the quincuncial arrangement of the setæ in the posterior region of the body; anteriorly the setæ are in couples.

The body-wall, being somewhat transparent, allowed me to distinguish through it paired light-coloured bodies, or "pyriform sacs," lying in the ventral region posteriorly. The most anterior somite of the body is very elongate, and carries the mouth terminally, the prostomium being absent. These features recalled the genus *Urochæta* of Perrier; but somewhat similar characters are found in other Earthworms: thus the pyriform sacs have been described by myself in *Urobenus* * ; and the scattered condition of the setæ, though not identical with the arrangement noticed, closely resembled, and might easily be confounded with, what obtains in *Diachæta* (Benham).

Turning then to the clitellum, I found it to cover the somites XIV. to XXII. or XXIII.; it is not complete on the ventral surface; and both anteriorly and posteriorly is, as is often the case, more feebly developed. One peculiar feature, however, about the clitellum, which therefore recalled *Urochæta*, is the fact that the intersegmental grooves are deep and noticeable; the glandular structure not being continuous from somite to somite, as is the case in most other Earthworms.

I could see no pores, or external apertures, of the genital ducts or nephridia; I therefore opened the worm, in the ordinary way, by a median dorsal incision, in order to satisfy my suspicion as to its belonging to the genus *Urochæta*.

The septa are thin, with the exception of four situated anteriorly, which are greatly thickened, namely those forming the posterior wall of somites VI., VII., VIII., and X. (the septum between IX. and X. is absent). Such thickened muscular septa are not unusual in Earthworms; but whether their position is constant in a given species is by no means certain. Perrier has not helped us to settle the matter, since in his figure he represents only four such septa, whereas in the text he speaks of five of them. However, Beddard †, in a species of this genus from Australia, describes four, having the same position as in the worm under consideration.

* Quart. Journ. Micr. Sci. vol. xxvii.
† Proc. Roy. Soc. Edinb. xiv. 1887, p. 160.

The alimentary tract, the vascular system, the nephridia, all exhibit the characters peculiar to or present in *Urochæta*.

There is but a single pair of seminal reservoirs, which have a greater extent than in *U. corethrura* (F. Müller *) : for in the present specimen that lying on the left side passes through eight somites, that of the right side passes through twelve somites, commencing in somite XII., where are situated also the ciliated rosettes.

I was unable to trace the sperm-ducts ; I could find no ovary ; I did not look for testes, as this would have necessitated some damage to the worm, which I was anxious to injure as little as possible.

There are three pairs of spermathecæ ; each is a very elongate, thin-walled sac, enlarged distally, and lying respectively in somites VII., VIII., IX. The chief difference between the two species of *Urochæta* that have received names lies in the different position of the spermathecæ. In *U. corethrura* they lie in somites VIII., IX., X. ; in *U. dubia* (Horst) they are found in somites VI., VII., VIII. ; in Beddard's specimen from Australia they have the same position as in the present specimen.

Such is a brief sketch of the anatomy of the worm A, from Fernando Noronha ; it is sufficient, however, to identify it as belonging to the genus *Urochæta* ; but as to the species— whether it belongs to any of those already described or requires a new name—I feel rather diffident of expressing an opinion. In most points it agrees closely with *U. corethrura* ; but in the position of the gizzard (in somite VI. instead of VII.), in the position of the spermathecæ, and in the fact that the setæ are not bifid, the two forms differ. On this last point I think no great stress can be laid, as Beddard recognizes no bifidity in his Australian specimen ; and I agree with him so far as the present specimen is concerned, which differs also from Horst's species, *U. dubia*, in the position of the spermathecæ ; in fact, with regard to these organs, the present and Beddard's specimen are intermediate between Horst's and Perrier's species. But are we justified in establishing a new species on such slender grounds, and from an examination of a single specimen ? I think not, and prefer to leave the specimen unnamed, and to regard it as belonging to Perrier's species, of which it may be a variety ; for we are at present ignorant as to how far

* See Perrier, Arch. d. Zool. expér. et gén. iii. 1874.

variation may occur in Earthworms; since with the exception of Beddard's paper on *Perionyx* (Journ. Linn. Soc., Zool. 1886, p. 308), we know absolutely nothing of the subject, and the present specimen forms a step between *U. corethrura* and *U. dubia.*

The two small worms resemble the specimen A in colour and in external characters; the clitellum is, however, undeveloped, so that they are probably young specimens of the same worm.

The worm B is longer than A, and of a somewhat different colour, being of a rather more reddish or violet-brown tint.

The body-wall is transparent, and showed white pyriform sacs through it much more distinctly than is the case with A.

The worm is, however, so soft that no setæ protrude, and I was unable to satisfy myself as to their exact arrangement posteriorly; anteriorly they are paired; posteriorly they are scattered, but whether regularly or not I cannot be positive.

The clitellum occupies somites XIV. to XXIX., and is thus rather more extensive than in A; but the worm is so soft that it is difficult to count with accuracy the somites, as some of the rings may be merely annuli. Thus far, then, we have no indication as to its genus; but on opening it, the arrangement of the septa, seminal reservoirs, and spermathecæ are seen to agree with what is found in A.

This second worm is therefore *Urochæta*, and doubtless the same species as the preceding.

It will be seen that I have made no morphological studies of these worms, nor sought to do more than identify them. Indeed, they were too badly preserved to be of any use histologically, and I should not have felt justified in sectionizing them even if they had been in good condition.

The fact that these worms belong to the genus *Urochæta*, which has been already described from Brazil and some of the neighbouring islands, lends considerable support to Mr. Ridley's supposition that they have been imported from the mainland in the mould in which cultivated plants were brought to the island. In conclusion I must express my best thanks to Mr. Ridley for allowing me to examine and identify them.

PORIFERA.

By H. J. CARTER, F.R.S.

Dry Specimens.

These were all too much beach-worn for specific distinction. The Nos. correspond with those on the Specimens.

1. POLYTHERSES, *Duchassaing et Mich.**

2. HIRCINIA.

3. CHALINA.—Spicules fine, slender, acerate.

4. HIRCINIA.—Fine structure.

5. EUSPONGIA ("best Turkey Sponge" of commerce).

6. EUSPONGIA.—Bearing *Polytrema miniaceum.*

7. HIRCINIA.—Skeletal structure partially filled with the filaments of *Spongiophaga communis.*

Wet Specimens.

Most of these are too fragmentary for specific distinction, although possessing the natural characters which they presented when taken from their habitat.

8. POLYTHERSES.—Two coarse pieces alone; the rest on pieces of a fine *Hircinia.*

9. EUSPONGIA ("best Turkey Sponge" of commerce").—Three or four discoloured pieces.

10. CHONDRILLA NUCULA, *Sdt.*

11. GEODIA.—? *G. Tumulosa, Bk.*—Siliceous balls spherical. Zone-spicule trifid; arms simple, undivided, extending upwards, outwards, and lastly horizontally. Bearing *Polytrema.*

12. CHONDROPSIS ARENIFERA, *Cart.* (Ann. & Mag. Nat. Hist. 1886, vol. xvii. p. 122).—Acuate spicules, sometimes blunt at each end.

---

* It should be remembered that "*Polytherses*" is a *Hircinia* in which the soft parts have been replaced by a structure composed of the filaments of *Spongiophaga communis*, Cart., which is of world-wide occurrence, but of which the nature is still unknown.

13. CHALINA? species.—Dark, dirty; fragments still bearing traces of their natural red-purple colour. Extending horizontally; throwing up thick ridges; scattered over with short, erect, tubular vents of different lengths. Fibre tough, charged abundantly with comparatively large, acerate spicules. Several pieces, some accompanied by a portion of *Chondropsis arenifera.*

14. EUSPONGIA ("fine Turkey sponge" of commerce).— Typically good, but small specimen; presenting the characteristic, crinkled surface. Colour black above, light sponge-yellow below. Bearing *Polytrema.*

15. CHONDRILLA NUCULA, *Sdt.*—Typically good specimen, growing over sand-detritus mixed with *Polytrema.*

16. CHONDRILLA PHYLLODES, *Sdt.*—Antilles. Spicules of two kinds, viz., pin-like skeletal, and spinispirular flesh-spicule. Closely allied in this respect to *Spirastrella cunctatrix,* Sdt. Colour grey or violet. Consistence gelatinous, firm. Three typically good specimens growing over sand-detritus mixed with *Polytrema* covered with white *Melobesia.*

17. CHONDROPSIS ARENIFERA.—Black on the surface from a layer of brown pigmental cells. Growing over a black *Stelletta* (? species), also bearing a cortical layer of dark brown-black pigmental cells mixed with small stellates. Zone-spicule trifid. Arms simple, straight, extended upwards and outwards.—Four large pieces.

18. CHONDRILLA NUCULA, *Sdt.*—Small, but typically good specimen.

19. SYNASCIDIA.—Common form. Globular, radiated calcareous spicule. Colour purple-white. Two pieces.

20. CHALINA? species.—Same as No. 13. One piece bearing a bit of *Hircinia.*

21. EUSPONGIA ("best Turkey sponge" of commerce).—Three small pieces.

22. ? ALCYONIUM or HYDROID ZOOPHYTE.—Digitate, reptant; colour yellowish; consistence soft.

23. ACTINIA ? sp.—Now lead-colour.

24. ALCYONIUM.—Congregated, short, columnar individuals; constricted circularly throughout the column.

25. EUSPONGIA ("Honeycomb sponge" of commerce).—Coarse cavernous structure.

26. EUSPONGIA.—Ditto.

27. GEODIA, same as No. 11.—Fragments of skin and body-substance only.

28. DONATIA LYNCURIUM.—Four specimens; the largest ¾ in. in diameter.

29. POLYTHERSES, with skeletal structure of *Hircinia* protruding.

30. EUSPONGIA ("Honeycomb sponge" of commerce).—Coarse cavernous structure.

31. SUBERITES MASSA, *Sdt.*—Spicule of one form only, simply pin-like, with subglobular head. Eight fragments. Surface warty. Colour yellowish.

32. AXINELLA ? species.—Form of specimen globular, about an inch in diameter, composed of radiating, erect, tough fibre, charged throughout with projecting tufts of simply acuate curved spicules; of one form only.

33. TETHYA CRANIUM.—Two small, discoloured, black fragments.

34. CHONDROPSIS ARENIFERA.—Specimen triangular, elongate; 4 inches long. Bearing *Polytrema* and *Melobesia*.

35. EUSPONGIA ("Honeycomb sponge" of commerce).—Coarse, cavernous structure. Three pieces. Colour black above, light sponge-yellow below.

36. CHONDROPSIS ARENIFERA.—Good, but small typical specimen.

37. ? HYMENIACIDON SANGUINEA, *Bk.*—Small insignificant specimen. Spicule of one from only, viz. pin-like and slightly curved. Colour now yellowish.

38. LEUCONIA SACCULATA, n. sp. (Calcisponge).—Form of specimen sacculated, consisting of four or more inflations

projecting from a common cavity. Colour white. Entire specimen about an inch long, and $\frac{3}{4}$ of an inch in diameter ($1 \times \frac{3}{4} \times \frac{3}{4}$ inch); broken out on one side, if not the point of attachment. The uppermost or principal division ending in a peristomatous mouth, which can only be seen with a microscope, hence to the unassisted eye looks "naked"; $\frac{2}{13}$ in. in diameter. The same on the summit of each inflation, but reduced to the size of a pin's head. Surface of the body smooth, composed of intercrossing arms of quadriradiate spicules *only*, between which are the pores. Vents as just described, leading into a general cloacal cavity corresponding in its inflations with those of the body. Surface of the cloaca scattered over with holes of very different sizes, very irregularly situated in a layer of minute quadriradiates whose fourth arm is much smaller than the rest, curved towards the oral orifice and projecting plentifully above the surface of the cloaca. The spicules of three kinds, viz. 3-radiate, 4-radiate, and linear cylindrical acerate.

Wall of the body about $\frac{1}{16}$ in. in diameter, composed of three layers of spicules, viz. :—1, consisting of comparatively large quadriradiates whose shaft projects inwards and whose other three arms are spread out horizontally over the surface; arm about $\frac{1}{45}$ by $\frac{1}{300}$ in. in its greatest diameters; 2 (the middle substance of the wall), consisting of 3- and 4-radiates mixed irregularly, whose arms are about the same size as that of the quadriradiates of the first or external layer; 3, or internal layer, forming the surface of the cloaca, and consisting of *minute* or infinitely smaller quadriradiates, whose shafts are directed outwards; the other two arms horizontal, and the fourth, or "spine" as it has been called, which is much the smallest, projecting above the surface in the way mentioned. Peristome consisting of palisading spicules about $\frac{1}{300}$ in. long, very fine and straight with abruptly pointed ends, crossed and kept in position by the spreading arms of the quadriradiates of the third, or cloacal, layer, here much enlarged. Wall permeated by branched canals, which commencing on the pores on the surface end in the holes or apertures on that of the cloaca.

Loc. Island of Fernando Noronha.

39. GEODIA.—Fragments of skin and body-substance. Same as No. 11, &c.

Summary of Specimens, arranged according to the Author's Classification (Ann. Mag. N. H. 1875, vol. xvi. p. 43).

Order I. CARNOSA.

Family GUMMINIA.

CHONDRILLA NUCULA, *Sdt.*   10, 15, 18.
CHONDRILLA PHYLLODES, *Sdt.*   16.

Order II. CERATINA.

None.

Order III. PSAMMONEMATA.

EUSPONGIA ("fine Turkey sponge" of commerce).   5, 6, 9, 14, 21.

EUSPONGIA ("Honeycomb sponge" of commerce).   25, 26, 30, 35.

HIRCINIA.   2, 4, 7.
POLYTHERSES.   1, 8, 29.

Order IV. RHAPHIDONEMATA.

CHALINA.   3, 13, 20.

Order V. ECHINONEMATA.

AXINELLA.   32.

Order VI. HOLORHAPHIDOTA.

HYMENIACIDON SANGUINEA, *Bk.*   37.
SUBERITES MASSA, *Sdt.*   31.
DONATIA LYNCURIUM.   28.
CHONDROPSIS ARENIFERA, *Curt.*   12, 17, 34, 36.
GEODIA,   11, 27, 39.
TETHYA CRANIUM.   33.

Order VII. HEXACTINELLIDA.

None.

Order VIII. CALCAREA.

LEUCONIA SACCULATA, n. sp.   38.

Remarks.

Of the present collection, it may be stated that the facies is West-Indian. *Chondrilla phyllodes*, Sdt., has as yet only been chronicled by that author, and that, too, from the "Antilles" (Grundzüge Spong.-F. atlantisch. Gebietes, p. 26). "*Poly-therses*" was the name given by Duchassaing and Michelotti to this transformed, sponge-like body which they dredged in the Caribbean Sea, and whose constituent parts have been above noticed. Both sorts of the Officinal Sponge, viz. the "Best Turkey" and the "Honeycomb" of Commerce, are plentiful, as they are in the West Indies generally. The Calcisponge, *Leuconia sacculata*, is a new species.

MADREPORARIA.

By Prof. P. MARTIN DUNCAN, M.B. (Lond.), F.R.S., &c.

Section MADREPORARIA APOROSA.

Family ASTRÆIDÆ, *Ed. & H.*

Subfamily ASTRÆIDÆ REPTANTES.

ASTRANGIA SOLITARIA, *Lesueur*, sp., *Verrill, Bull. Mus. Comp. Zoöl. No. 3; Pourtalès,* 1871, *Ill. Cat. Mus. Comp. Zoöl.* p. 79.

A specimen much covered with Nullipores. The corallites smaller than the Florida types, and the so-called "pali" not united before the second and third cycles of septa. The form is allied to *Astrangia Danæ* and to *A. Michelini.* It may be considered as a small variety of the Floridan species.

Subfamily ASTRÆIDÆ CÆSPITOSÆ.

MUSSA, sp.—A worn specimen of a species.

Subfamily ASTRÆIDÆ AGGLOMERATÆ FISSIPARANTES.

FAVIA CONFERTA, *Verrill,* 1867, *Trans. Connect. Acad.* vol. i., in *Hartt's Collection of Corals from the Abrolhos Reef, Brazil,* p. 355 (1868).

This species is remarkable for its elongated calices due to fissiparity, and for their closeness. Often a ridge separates the neighbouring calices, instead of some vestige of interspace and costæ.

Verrill and Pourtalès noticed the alliance of the species with the genera *Goniastræa* and *Mæandrina*. In a small specimen the Goniastroid appearance is striking, and the costæ between the calices are only visible at one spot. There are several rolled specimens, and, as Pourtalès remarked, they resemble Goniastroids very much.

FAVIA ANANAS, *Lamarck*, sp.

The specimens have the usual well-developed columella, the costæ are very visible in one, and the fourth cycle of septa is incomplete.

FAVIA DEFORMATA, *Ed. & H., Hist. Nat. des Corall.* vol. ii. p. 431.

A somewhat worn and broken coral, very Cœlorian in appearance, and with large calices, some long and serial in appearance, but really the result of fissiparity, appears to be a specimen of this form, the habitat of which has hitherto been unknown. The walls are close, but in many places their former separation can be shown. It is a very erratic species, and better specimens are required.

A very worn specimen of the same species was also obtained.

These two specimens came from Rat Island. They were washed up on the south-east corner with a number of Sponges.

Section MADREPORARIA FUNGIDA.

Family PLESIOFUNGIDÆ, *Duncan*.

Revision of the Genera of Madreporaria, Journ. Linn. Soc. vol. xviii. 1884, p. 133.

SIDERASTRÆA SIDEREA, *Ellis & Soland.*, sp.

This specimen, and indeed all the rolled ones besides, have smaller calices than the Caribbean type, but that is the only distinction. They greatly resemble *Siderastræa stellata*, Verrill, from the Abrolhos Reef.

Remarks.

This little Coral fauna has the Abrolhos Reef homotaxis, and the species are fairly intermediate between those of the same genera of that reef and of the Caribbean Sea.

ERRATA.

Page 446, line 19 from bottom, for *Centronotus gunellus* read *Centronotus gunnellus*.

474, line 12 from bottom, for *Pompilus nesophila* read *Pompilus nesophilus*.

503, line 9 from bottom, for *Chiton (Ischnochiton) carribæorum* read *Chiton (Ischnochiton) caribbæorum*.

555, line 7 from top, for *A. jaspicea* read *A. jaspidea*.

www.ingramcontent.com/pod-product-compliance
Lightning Source LLC
Chambersburg PA
CBHW021707210326
41599CB00013B/1559